Evans Objective Tests in Mathematics

Murray Macrae B.Sc.(Edin.)
P.G.C.E.(Lond.)

Evans Brothers Limited

Published by Evans Brothers Limited
Montague House, Russell Square,
London WC1B 5BX

Evans Brothers (Nigeria Publishers) Limited
PMB 5164, Jericho Road
Ibadan

© M. F. Macrae 1970
First published 1970
Second impression 1972

All Rights Reserved. No part of this publication may be reproduced, stored in a retrieval system, or transmitted, in any form or by any means, electronic, mechanical, photocopying, recording or otherwise, without the prior permission of Evans Brothers Limited.

Illustrations by Murray Macrae and V. Taylor

Printed in Great Britain by William Clowes & Sons, Limited, London, Beccles and Colchester
ISBN 0 237 28758 7   PRA 2583

Contents

Preface		page v
Notation		vi

Part I, Class Topics		1
Chapter 1	Area and Volume	2
Chapter 2	Arithmetic Conversion	14
Chapter 3	Arithmetic in Different Number Bases	22
Chapter 4	Arithmetic: Rate, Ratio, Percentage	33
Chapter 5	Geometry and Trigonometry in Three Dimensions	42
Chapter 6	Graphs: Interpretative and Algebraic	56
Chapter 7	Latitude and Longitude	69
Chapter 8	Linear Inequalities and Undefined Fractions	81
Chapter 9	Plan and Elevation	91
Chapter 10	Plane Geometry	111
Chapter 11	Plane Trigonometry	123
Chapter 12	Velocity Triangles	133

Part II, Certificate-level Tests	149
Test 1	150
Test 2	166
Test 3	184
Test 4	202
Test 5	218
Test 6	234

Preface

This book has been written to provide students and teachers with a wealth of examples of objective or multiple choice questions at about the level of the West African School Certificate.

The book is in two parts:

Part I contains twelve class tests of twenty questions. Each test covers a separate topic and is designed to last forty minutes.

Part II contains six tests of fifty questions. These tests cover the full range of the normal Syllabus B Mathematics course and each is designed to last 1¼ hours.

The questions, which were devised in order to prepare IVth and Vth form students for the West African School Certificate, have been tried and tested in the classroom. The distractors (incorrect answers) contain the likely errors.

I would like to thank the following for their help in the preparation of this book. My colleagues, Mohammed Mamman Balkore and Val Montgomerie, who assisted me in teaching the material in class. The 1968 and 1969 'finalists' of Government Secondary School, Bida. As a result of their co-operation I was able to make useful modifications to the original material. My wife, Lynne, who made many valuable suggestions while the manuscript was being prepared and who checked the solutions to the questions.

M.F.M.

Notation

The system of notation used in this book follows that which was recently introduced by the West African Examinations Council. The less common notations are described below.

Notation	Means
$\simeq$	is approximately equal to
$\neq$	is not equal to
AB	the line AB
$\overline{AB}$	the line segment AB
|AB|	the distance, or length of, AB
$X\hat{Y}Z$	angle XYZ
$\triangle ABC \equiv \triangle XYZ$	triangle ABC is congruent to triangle XYZ
$\triangle ABC \;\|\|\|\; \triangle XYZ$	triangle ABC is similar to triangle XYZ
$\triangle ABC = \triangle XYZ$	triangle ABC is equal in area to triangle XYZ
AB $\|\|$ XY	AB is parallel to XY
AB $\perp$ XY	AB is perpendicular to XY

Part I: Class Topics

The topics in Chapters 1-12 have been chosen in the light of three main considerations:
(i) Those topics which students generally find difficult.
(ii) Recent additions to the West African School Certificate B Mathematics syllabus.
(iii) Topics which involve work in three dimensions with an emphasis on visual problems.

Each test is preceded by a short note which includes:
(i) A note on what the student should know of the topic before attempting the test.
(ii) Some advice on solving problems in the topic.

The tests are graded, the more difficult questions appearing at the end of the test, and they should be done in the normal lesson time of forty minutes.

Chapter 1 Area and Volume

For Certificate purposes the student is required to know how to calculate, from given dimensions, the areas of certain plane figures: the triangle; the quadrilaterals; the polygons; the circle. In addition, he should know how to calculate the areas of certain curved surfaces: the curved surface of a cylinder; the curved surface of a cone; the surface area of a sphere.

The student should also know how to calculate, from given dimensions, the volumes of the following: solids of uniform cross-section (the cube, the cuboid, the cylinder, the prisms); pyramids (right cone, square based pyramid); the sphere.

Below is a summary of the more common question types encountered in problems on area and volume:

1. By Direct Substitution into Formula

In this type of question the dimensions are given, or can be found from the information in the question, and the solution is obtained by substituting these into the correct formula. For example:

'What is the volume of a cone whose radius is $3\frac{1}{2}''$ and whose height is $6''$? (Take $\frac{22}{7}$ for π)'

In this case, $\qquad V = \frac{1}{3}\pi r^2 h \qquad$ (standard notation)

therefore, substituting the given dimensions,

$$V = \tfrac{1}{3} \times \tfrac{22}{7} \times \tfrac{7}{2} \times \tfrac{7}{2} \times \tfrac{6}{1} \text{ cub. in.}$$

$$V = 77 \text{ cub. in.}$$

However, the problem is not always as simple as this. What if the question had been stated as follows?

'What is the radius of a cone whose height is 6" and whose volume is 77 cub. in.? (Take $\frac{22}{7}$ for π)'

Here the same formula is used,

$$V = \tfrac{1}{3}\pi r^2 h \qquad \text{(standard notation)}$$

and, again, the dimensions should be substituted,

$$77 = \tfrac{1}{3} \times \tfrac{22}{7} \times r^2 \times \tfrac{6}{1}$$

both sides of the equation can now be divided by 11 and the right hand side of the equation can be simplified by dividing the 3 into the 6,

$$7 = \tfrac{2}{7} \times r^2 \times \tfrac{2}{1}$$

multiplying both sides of the equation by $\tfrac{7}{4}$,

$$\tfrac{49}{4} = r^2$$

taking the square root of both sides of the equation,

$$r = \tfrac{7}{2} \text{ in.}$$

The student should be familiar with this type of operation for the various formulae he knows.

2. Addition and Subtraction Methods

Often a question will give a figure or a solid which has been obtained by adding or subtracting more simple figures (or solids). For example:

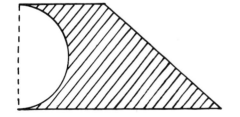

A semicircle has been cut out of a trapezium.

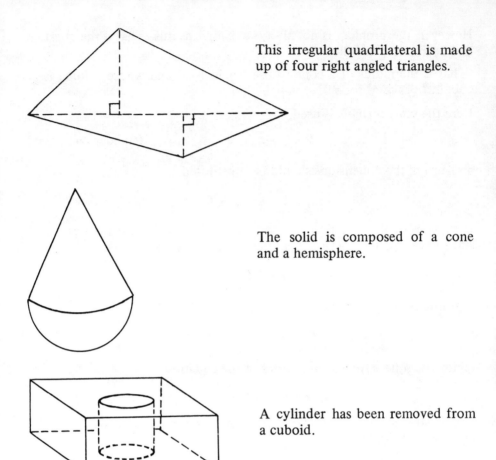

This irregular quadrilateral is made up of four right angled triangles.

The solid is composed of a cone and a hemisphere.

A cylinder has been removed from a cuboid.

3. Calculation of Area by Trigonometrical Methods
If a question on area contains information about an angle, it is likely that the question has to be solved by a trigonometrical method. In particular:

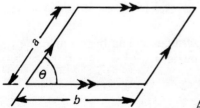

Area of the given parallelogram = $ab \sin \theta$.

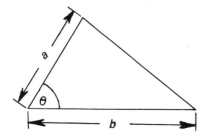

The area of the given triangle = $\tfrac{1}{2} ab \sin \theta$.

4. Transformation of Solids

The student will find that, in certain questions, a given solid will undergo a change of shape. For example:

(a) 'A cube of ice melts and the water is collected in a cylindrical beaker. . . .'
(b) 'A solid metal cone is melted down and recast as a cuboid. . . .'
(c) 'A cylindrical tin of evaporated milk will just fill a hemispherical cup. . . .'

In each case, the *shape* of the orginal solid has been *changed*, but the *volume* has remained *unchanged* during the transformation. This knowledge should be used when solving the problem.

Class Test: Area and Volume
40 minutes 20 questions

1. The area of △ABC is 8 sq. cm., |AB| = 4 cm. and |AC| = 5 cm. What is BÂC?

    A. 126° 52'

    B. 141° 20'

    C. 143° 8'

    D. 156° 25'

    E. None of the above

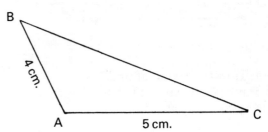

2. If, in addition to the information of the previous question, you are told that |BC| = 8·062 cm., what is the length of the perpendicular from A to $\overline{BC}$?

    A. 0·5045 cm.

    B. 0·9259 cm.

    C. 1·008 cm.

    D. 1·985 cm.

    E. 2·012 cm.

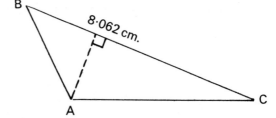

3. What is the volume, to the nearest cubic inch, of a cylinder whose axis is 10 inches long and whose cross-sectional diameter is 3 inches? (Approximate π to $\frac{22}{7}$)

    A. 283 cub. in.

    B. 141 cub. in.

    C. 110 cub. in.

    D. 94 cub. in.

    E. 71 cub. in.

4. Three boys were asked how they would find the area of the irregular quadrilateral shown. Here is what they replied:

Abdul: Draw $\overline{AC}$; from B and D draw lines to meet $\overline{AC}$ at right angles. The areas of the four right angled triangles can be calculated after suitable measurement and thus the area of the whole figure can be found by addition.
Baba: Draw $\overline{BD}$. By measuring any two sides and the included angle of the two triangles formed, their areas can be calculated. Add these to get the area of the quadrilateral.
Chinue: Use the formula.

Which boy(s) was (were) correct?

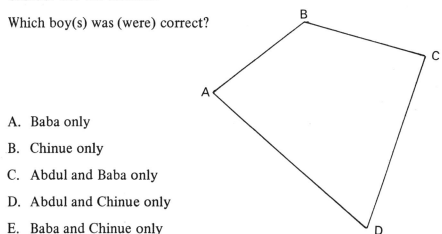

A. Baba only

B. Chinue only

C. Abdul and Baba only

D. Abdul and Chinue only

E. Baba and Chinue only

5. In the diagram, $\overline{AF} \parallel \overline{DC}$ and $\overline{AD} \parallel \overline{BC}$. X is a point on $\overline{DC}$. What is the area of rectangle EFCD if $\triangle ABX$ = 2 sq. in.?

A. 1 sq. in.

B. 2 sq. in.

C. 4 sq. in.

D. 6 sq. in.

E. 8 sq. in.

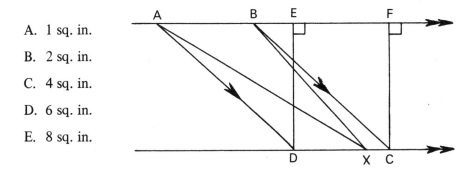

6. A wedge is in the form of a right prism of triangular section. The base of the wedge is a rectangle, 1 ft. by 8 in. and the highest edge of the wedge is 3 in. above the base. What is the volume of the wedge?

A. 288 cub. in.
B. 144 cub. in.
C. 96 cub. in.
D. 72 cub. in.
E. 12 cub. in.

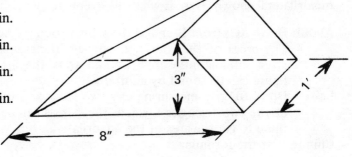

7. What is the volume of a cuboid which is twice as wide as it is high, and whose height is one-and-a-half times its length, l?

   A. $9l^3$
   B. $4\frac{1}{2}l^3$
   C. $3\frac{1}{2}l^3$
   D. $3l^3$
   E. l^3

8. The design shown, known as the Monad, is a religious symbol of the Orient. It is constructed by drawing two semicircles on a diameter of the outer circle. If the radius of the outer circle is $2r$, what is the area of the shaded portion of the Monad?

   A. πr^2
   B. $\frac{3}{2}\pi r^2$
   C. $2\pi r^2$
   D. $3\pi r^2$
   E. $4\pi r^2$

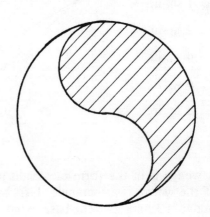

9. What is the area of a square whose diagonal is 14 in. in length?

   A. 7 sq. in.
   B. 28 sq. in.
   C. 49 sq. in.
   D. 98 sq. in.
   E. 196 sq. in.

10. In the diagram, OAB is a sector of a circle, radius 14 in. AÔB = 45° and $\overline{AD} \perp \overline{OB}$. What is the area of ABD? (Take for π.)

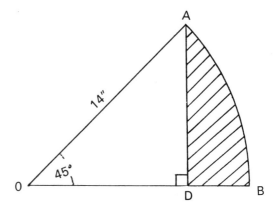

   A. 77 sq. in.
   B. 49 sq. in.
   C. 38·5 sq. in.
   D. 28 sq. in.
   E. 21 sq. in.

11. The diagram on the right shows a view of a right pyramid. It has five faces—four equilateral triangles and a square.

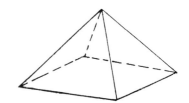

From which of the following plane figures could such a pyramid be constructed by folding along the dotted lines only? (Each figure contains four equilateral triangles and one square.)

A.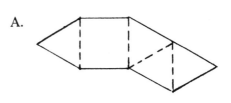

B.

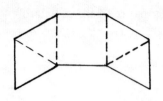

C.

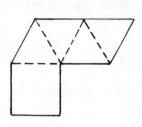

D.

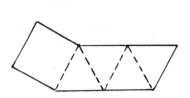

E.

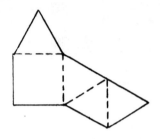

12. In the diagram, $B\widehat{M}A = C\widehat{N}A = 1$ right angle. $|AM| = |MN| = |BX| = 4"$, $|CN| = 6"$ and $|XM| = 3"$. What is the area of $\triangle ABC$?

A. 26 sq. in.

B. 24 sq. in.

C. 20 sq. in.

D. 16 sq. in.

E. 14 sq. in.

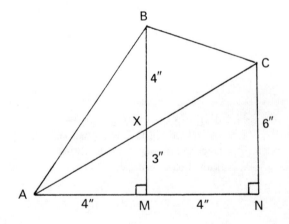

13. A cup is in the form of an inverted cone of depth d and base-radius r. If the water which just fills a hemispherical bowl of radius r is poured into the cup its level is a distance $\frac{1}{2}d$ above the vertex of the cup. What is d in terms of r? (Vol. of a sphere, rad. r, is $\frac{4}{3}\pi r^3$.)

10

A. $d = 32r$
B. $d = 16r$
C. $d = 8r$
D. $d = 4r$
E. $d = 2r$

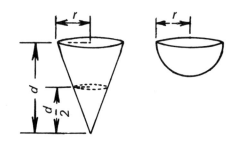

14. A right pyramid has a square base ABCD. Its vertex, V, is 12 in. above the base. If |AB| = 10 in., what is the area of the inclined face VAB?

A. 50 sq. in.
B. 55 sq. in.
C. 60 sq. in.
D. 65 sq. in.
E. 70 sq. in.

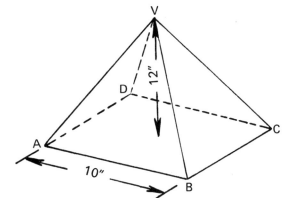

15. A vee-block is a device used in an engineering workshop for holding metal plates at right angles to each other. Thus, in the diagram, $A\hat{V}B$ is a right angle. Using the dimensions and information contained in the diagram, calculate the volume of the block.

A. 34 cub. in.
B. 112 cub. in.
C. 272 cub. in.
D. 328 cub. in.
E. 336 cub. in.

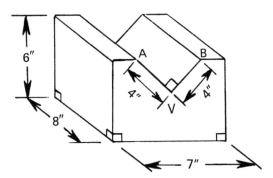

16. An open-ended paper cone has a slant height of 7 in. and a base radius of 3½ in. If the cone is cut along a straight line from a point P on the base to its vertex, V, and the paper is flattened out, what will be the area of the paper? (Approximate π to $\frac{22}{7}$.)

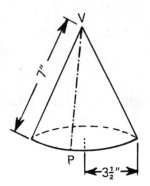

A. 77 sq. in.

B. 51⅓ sq. in.

C. 38½ sq. in.

D. 25⅔ sq. in.

E. 14 sq. in.

17. In the diagram, equilateral △ABC is shown, also its inscribed circle, radius r, and its circumcircle, radius R. From the construction indicated on the diagram, find R in terms of r, and use this to express the area of the shaded ring in terms of r.

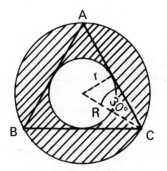

A. $\tfrac{1}{3}\pi r^2$

B. $\tfrac{3}{4}\pi r^2$

C. πr^2

D. $2\pi r^2$

E. $3\pi r^2$

18. A spherical balloon initially contains V cc. of gas. The balloon develops a small hole which allows the gas to escape slowly. When the size of the balloon has decreased so that its original radius has been halved, what volume of gas, in terms of V, has escaped? (The volume of a sphere is directly proportional to the cube of its radius.)

A. $\tfrac{7}{8}V$

B. $\tfrac{3}{4}V$

C. $\tfrac{1}{2}V$

D. $\tfrac{1}{3}V$

E. $\tfrac{1}{8}V$

19. A plastic measuring spoon is given away free with every tin of milk powder. The spoon holds 8·8 cc. and 250 spoonfuls can be obtained from a full tin. If the tin is in the shape of a cylinder of diameter 14 cm., what is its height? (Take π to be $\tfrac{22}{7}$)

    A. 50 cm.

    B. 14·3 cm.

    C. 14 cm.

    D. 7·1 cm.

    E. 3·8 cm.

20. The diagram shows a view of a cylindrical vessel which has been cut by a plane of symmetry. The external radius and height of the vessel are R and h respectively and the thickness of the walls is t. Which of the following is an expression for the internal volume of the vessel?

    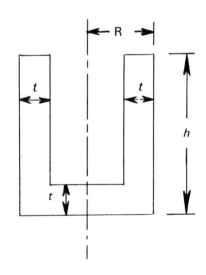

    A. $\pi R(Rh - \tfrac{1}{2}t^2)$

    B. $\pi R^2 h - \pi t^3$

    C. $\pi R^2 h - \pi t^2(h - t)$

    D. $\pi(R - 2t)^2(h - t)$

    E. $\pi(R - t)^2(h - t)$

Chapter 2 Arithmetic Conversion

The student should know, and have worked with, the British and metric systems of measurement of length, weight and cubic capacity. In addition, he should be familiar with the currencies of West Africa, Britain and the United States.

Much of the difficulty met with in this topic arises because many students think that there is a formula to cover each question. As there is a great variety of measures which can be converted to other related measures, it is better to find a method of working which can be used in the various situations than to rely upon the use of a formula which very likely does not exist in any case.

The following examples have been worked out using the same basic method in each case:

(a) 'If £N1 = N₵2.85, how many Nigerian pounds would be given in exchange for N₵57.00?'

Start by writing down the given conversion rate,
$$£N1 = N₵2.85$$

Multiply both sides of the equation by $\dfrac{57.00}{2.85}$,

$$£N1 \times \dfrac{57.00}{2.85} = N₵2.85 \times \dfrac{57.00}{2.85}$$

$$£N \dfrac{57.00}{2.85} = N₵57.00$$

$$£N20 = N₵57.00$$

Thus, £N20 would be given in exchange for N₵57.00.

(b) 'If an aircraft travels at 660 m.p.h., how many feet does it travel in one second?'

No rate of conversion is given but we know that
$$1 \text{ mile} = 3 \times 1760 \text{ feet}$$
Therefore, $\quad$ 1 m.p.h. = 3×1760 feet/hour
Multiplying both sides of the equation by 660,
$$660 \text{ m.p.h.} = 3 \times 1760 \times 660 \text{ feet/hour}$$
Distance travelled/hour = $3 \times 1760 \times 660$ feet
$$\text{Distance travelled/sec.} = \frac{3 \times 1760 \times 660}{60 \times 60} \text{ feet}$$
$$= 968 \text{ feet}$$
The aircraft travels 968 feet in one second.

(c) 'If 1 km. is approximately equivalent to 0·6214 mi., express 1 metre in miles.'

From the question,
$$1 \text{ km.} \simeq 0\cdot 6214 \text{ mi.}$$
Dividing both sides by 1000,
$$\frac{1}{1000} \text{ km.} \simeq \frac{0\cdot 6214}{1000} \text{ mi.}$$
$$1 \text{ metre} \simeq 0\cdot 0006214 \text{ mi.}$$
1 metre is approximately equivalent to 0·0006214 mi.

Method
(1) Write down the conversion rate. This will either be (a) given in the question, or (b) something that the student is expected to know.
(2) Leave all numerical simplification until the last step.

Class Test: Arithmetic Conversion
40 minutes 20 questions

1. If 6d. (Nigeria) can be exchanged for 17 francs (French-speaking countries), how many francs would be given in exchange for £N1?

    A. 4080 fr.

    B.   680 fr.

    C.   340 fr.

    D.   102 fr.

    E.   85 fr.

2. If 1 litre is equivalent to 1·75 pints, express 1 gallon in litres.

    A. 14 litres

    B. 4·57 litres

    C. 3·5 litres

    D. 0·22 litres

    E. 0·071 litres

3. A fan is rotating at 300 revolutions per minute. Through what angle does a blade turn in 1 second?

    A.   5°

    B.   72°

    C.   450°

    D.   600°

    E. 1800°

4. £N1 can be exchanged for Le2.00. How many cents (Sierra Leone) would be given in exchange for 1 shilling?

A. 0·1 c.
B. 0·5 c.
C. 10 c.
D. $16\frac{2}{3}$ c.
E. 50 c.

5. Given that 1 are is 100 sq. meters, what is 1 sq. cm. expressed in ares?

   A. 0·000001 ares
   B. 0·00001 ares
   C. 0·0001 ares
   D. 10,000 ares
   E. 1,000,000 ares

6. If 28·33 litres of water will just fill a container whose volume is 1 cubic foot, how many litres will fill a container whose volume is 1 cubic yard?

   A. 3·15 litres
   B. 9·44 litres
   C. 84·99 litres
   D. 254·97 litres
   E. 764·91 litres

7. A boy has a N₵ (New Cedis) and b Np (New pesewas). What is this amount expressed in New pesewas?

   A. $100ab$ Np
   B. $\frac{a}{100} + b$ Np

C. $a + \dfrac{b}{100}$ Np

D. $a + 100b$ Np

E. $100a + b$ Np

8. If 8 kilometres is the same distance as 5 miles, what is 1 metre expressed as a fraction of a mile?

   A. $\dfrac{1}{1600}$

   B. $\dfrac{1}{625}$

   C. $\dfrac{1}{160}$

   D. $\dfrac{10}{625}$

   E. $\dfrac{5}{8}$

9. If 1 gallon of liquid occupies 280 cub. in., how many gallons of water would be contained by a tank in the form of a cuboid, 8 in. by 1 ft. 2 in. by 2 ft. 1 in.?

   A. $\dfrac{1}{10}$ gal.

   B. $\dfrac{5}{8}$ gal.

   C. $7\dfrac{1}{2}$ gal.

   D. 10 gal.

   E. 120 gal.

10. An airline allows each passenger to carry 20 kilograms of luggage. If a passenger leaves Kano with 50 lb. of luggage, how many kilograms is he overweight? (1 kg. $\simeq$ 2·2 lb.)

    A. 90 kg.

    B. 13·6 kg.

    C. 13·2 kg.

D. 6 kg.

E. 2·7 kg.

11. If Le 2.00 can be exchanged for £N1, how many Nigerian shillings would be given for Lex?

   A. $\dfrac{x}{20}$

   B. $\dfrac{x}{10}$

   C. $\dfrac{10}{x}$

   D. $10x$

   E. $40x$

12. A motorist, travelling at 60 m.p.h., was suddenly dazzled by the headlights of an on-coming car. If 1 second elapsed before the other driver dipped his lights, how far did the motorist drive without being able to see properly?

   A. 5280 ft.

   B. 1056 ft.

   C. 88 ft.

   D. $29\frac{1}{3}$ ft.

   E. $1\frac{2}{3}$ ft.

13. If it costs £N15 in Lagos to fly from Lagos to Accra, how much would the return journey cost in Accra? (£N1 ≃ NC2.85)

   A. N₵42.75

   B. N₵30.35

   C. N₵29.93

   D. N₵19.00

   E. N₵ 5.26

14. A boy weighs 10st. 3lb. What is his weight in kilograms if 1 kg. is equivalent to 2·2lb?

   A.   13 kg.

   B.   47 kg.

   C.   65 kg.

   D.   74 kg.

   E.   314·75 kg.

15. A man runs 100 yards in 12·5 seconds. If he was able to keep this rate up for 1 mile he would break the world mile record. What would his time be?

   A.   2 min. 50 sec.

   B.   2 min. 54 sec.

   C.   3 min. 32·5 sec.

   D.   3 min. 40 sec.

   E.   3 min. 45 sec.

16. A hotel normally charges £N7 a day, but the manager lets Americans pay $19 a day instead. If the bank gives £N1 for $2.80, how much does an American save per day if he pays in dollars?

   A.   0

   B.   6 c.

   C.   40 c.

   D.   60 c.

   E.   96 c.

17. A Volkswagen's petrol tank holds 8·8 gallons. In Germany, where these cars are manufactured, capacity is measured in litres. If 1 gallon $\simeq$ 4·55 litres, what is the petrol tank's capacity, to the nearest litre?

   A.   45 litres

B. 40 litres

C. 36 litres

D. 32 litres

E. 2 litres

18. The length of the side of a square is 10 cm. What is the area of the square in square inches, if 1 inch is taken to be equal to 2·5 cm.?

    A. 625 sq. in.
    B. 250 sq. in.
    C. 40 sq. in.
    D. 25 sq. in.
    E. 16 sq. in.

19. After Sterling was devalued in November 1967, £N1 could be exchanged for £1 3s 4d (Sterling). How much Nigerian money, to the nearest penny, was given in exchange for £1 Sterling?

    A. 16s. 8d.
    B. 17s. 2d.
    C. £N1
    D. £N1 3s. 4d.
    E. £N1 7s. 3d.

20. An expatriate sends £N100 to his bank account in Britain. If the ratio of the values of £N1 to £1 Sterling is 7 to 6, how much, to the nearest £ Sterling, is credited to his account in Britain?

    A. £117
    B. £114
    C. £100
    D. £86
    E. £83

Chapter 3 Arithmetic in Different Number Bases

The student must be able to add, subtract, multiply and divide numbers in different bases. Also, he must be able to convert numbers from one base to another. The most common number bases are *two, three, five, eight, ten* and *twelve,* but practice must be gained in working with numbers of any reasonable base.

Number Structure
For a thorough understanding of this topic, it is essential to realize the significance of the position of the digits in a number. Thus:

$4121_{five} = (4 \times 5^3) + (1 \times 5^2) + (2 \times 5^1) + 1$
$279_{ten} = (2 \times 10^2) + (7 \times 10^1) + 9$
$11101_{two} = (1 \times 2^4) + (1 \times 2^3) + (1 \times 2^2) + (0 \times 2^1) + 1$
$6568_{nine} = (6 \times 9^3) + (5 \times 9^2) + (6 \times 9^1) + 8$

(Note that all the numbers on the right hand sides of the above equations have been written in base ten.)

Conversion Between Bases
(a) From any number base to base ten
The above knowledge should be put to use; thus, converting 510_{eight} to base ten:

$510_{eight} = (5 \times 8^2) + (1 \times 8^1) + 0$
$= 320_{ten} + 8_{ten} + 0$
$510_{eight} = 328_{ten}$

(b) From base ten to any other base
A quick method for changing a base ten number to a number in another base is shown in the examples below.
   Divide the base ten number by the number of the required base, writing down any remainder (even if it is zero). Repeat this process until there is nothing left to divide. The number in the required base is obtained by writing down the remainders in the opposite order to which they were found.

Converting 57_{ten} to base two.

2	57	Remainder
2	28	1
2	14	0
2	7	0
2	3	1
2	1	1
	0	1

Thus, $57_{ten} = 111001_{two}$

Converting 19340_{ten} to base twelve.

12	19340	Remainder
12	1611	8
12	134	3
12	11	2
	0	11

Let e represent the digit *eleven* in base twelve.
Thus, $19340_{ten} = e238_{twelve}$

(c) Conversions between bases other than base ten
Change 253_{seven} to a base four number.
   The most direct approach is to change 253_{seven} to a base *ten* number by method (a); then, using method (b), change the base ten number to base four:

$253_{seven} = (2 \times 7^2) + (5 \times 7^1) + 3$

$$= 98_{ten} + 35_{ten} + 3_{ten}$$
$$= 136_{ten}$$

		Remainder
4	136	
4	34	0 ↑
4'	8	2
4	2	0
	0	2

Thus, $253_{seven} = 2020_{four}$.

Addition
When adding a column of digits in base ten, 1 is carried into the next column each time the total of the digits reaches *ten*. The digit remaining is written down and the process is repeated in the next column, remembering to add any '1's which have been carried.

This method is extended to addition in other bases, except that 1 is carried when the digit total reaches the number of the base.

Consider the digits in the three columns of the following base eight addition:

$$\begin{array}{r} 353 \\ +267 \\ \hline 642 \end{array}$$

'units' column : $3 + 7 = 1 \times eight + 2$
'eights' column: $5 + 6 + 1$ (carried) $= 1 \times eight + 4$
'eight^2' column: $3 + 2 + 1$ (carried) $= 6$

Subtraction
Difficulty is met in subtraction when it is necessary to 'borrow' from the next column. In base ten, we borrow 1 *ten* from the top digit of the next column. Similarly, in base eight, 1 *eight* is borrowed; in base two, 1 *two* is borrowed; etc.

Consider the digits in the columns of the following base five subtraction:

$$\begin{array}{r} 322 \\ -134 \\ \hline 133 \end{array}$$

'units' column: 4 cannot be subtracted from 2, so 1 *five* is borrowed from the 2 in the next column. The 4 is now subtracted from 2 + *five* (*seven*), leaving . . . 3

'fives' column: 3 cannot be subtracted from the 1 remaining, so 1 *five* is borrowed from the 3 in the next column. The 3 is now subtracted from 1 + *five* (*six*), leaving . . . 3

'five^2' column: The 1 is subtracted from the 2 remaining, leaving 1.

Subtraction should always be checked by adding the bottom two rows of figures; their sum should equal the number at the top. Thus:

$$\begin{array}{r} 134 \\ +133 \\ \hline 322 \end{array}$$

Multiplication

The method of multiplying is the same for numbers of any base. The actual calculation depends on knowledge of the base ten multiplication tables. Any two digits are multiplied mentally in base ten and the resulting base ten number is changed to the required base. For example:

'What is the product, in base six, of 5_{six} and 4_{six}?'

(Mentally: $5 \times 4 = 20_{ten}$, but $20_{ten} = 32_{six}$)

$5_{six} \times 4_{six} = 32_{six}$

Check that this method has been used in the longer base six multiplication which follows:

$$\begin{array}{r} 4225 \\ 54 \\ \hline 25352 \\ 340210 \\ \hline 410002 \end{array}$$

Division

Again, the method of dividing is the same for numbers of any base and again the calculation depends on the use of the base ten multiplication tables. Thus, work the division mentally in base ten, remembering to write down the results in the required base.

Check each stage of the following base eight division:

$$\begin{array}{r} 45 \\ 5\overline{)271} \\ 24 \\ \hline 31 \\ 31 \\ \hline 0 \end{array}$$

Alternative Names for Numbers in Certain Bases

Base ten numbers are said to be in the *scale* of ten, similarly, base two numbers are said to be in the *scale* of two, and so on. Some of the scales have special names; these are listed below:

Base two:	Binary scale
Base three:	Ternary scale
Base eight:	Octal scale
Base ten:	Denary, or Decimal, scale
Base twelve:	Duodecimal scale

Working in Bases other than Ten

It is preferable, and ultimately quicker, to do any working in the base given in the question.

If, however, the student gets 'stuck' on a problem, he may find it easier to change the given numbers to base ten, work the problem in that base and convert the result back to the required base.

Class Test: Arithmetic in Different Number Bases

40 minutes 20 questions

Note (1) In some questions the number base is indicated by a printed suffix after the digits of the number. For example:

75_{eight} is the base eight number 75.

(2) The letters t and e are used to represent the base twelve digits *ten* and *eleven*.

1. If 546_{seven} is equivalent to $(5 \times P) + (4 \times Q) + (6 \times R)$, what do the letters P, Q and R represent?

   A. 100, 10 and 1 respectively

   B. 7^3, 7^2, and 7^1 respectively

   C. 7^2, 7^1 and 1 respectively

   D. 14, 7 and 1 respectively

   E. 700, 70 and 7 respectively

2. What base five number is equivalent to $(4 \times 5^3) + (0 \times 5^2) + (2 \times 5^1) + (2 \times 5^0)$?

   A. 512

   B. 3210

   C. 402

   D. 422

   E. 4022

3. What is the decimal number 29 expressed as a binary number?

   A. 11101

   B. 11111

C. 11011

D. 111101

E. 10011

4. What is the base twelve number *t0e* expressed as a decimal number?

   A.   21

   B.   131

   C.   1011

   D.   1451

   E.   10011

5. What is the binary number 10001 expressed as a decimal number?

   A.   2

   B.   5

   C.   9

   D.   15

   E.   17

6. What is the sum of the base six numbers 5043 and 4444?

   A. 111531_{six}

   B. 13531_{six}

   C. 12858_{six}

   D. 9487_{six}

   E. 2143_{six}

7. What is the product, expressed as a binary number, of 1001_{two} and 110_{two}

   A.   1111

B. 11011

C. 101101

D. 110110

E. 1001110

8. What is the value of x in the following equation?
$$2202_{five} = 1014_{five} + x_{five}$$

   A. 1188

   B. 1143

   C. 1133

   D. 1043

   E. 168

9. What is 150_{twelve} divided by 2_{twelve}?

   A. 75_{twelve}

   B. 86_{twelve}

   C. 89_{twelve}

   D. $t2_{twelve}$

   E. $e2_{twelve}$

10. If the sum of 6113 and 5164 is 12277, in what base are the numbers?

   A. eight

   B. nine

   C. ten

   D. eleven

   E. twelve

29

11. If 251 x 4 = 1124, in what base are the numbers?

    A. six

    B. seven

    C. eight

    D. nine

    E. ten

12. If the following subtraction is correct, in what base are the numbers?

    $$\begin{array}{r} 6153 \\ -2441 \\ \hline 3512 \end{array}$$

    A. eight

    B. nine

    C. ten

    D. twelve

    E. seven

13. If 101 is obtained from dividing 1010 by 10, in what base are the numbers?

    A. Base two only

    B. Base five only

    C Base eight only

    D. Base ten only

    E. The numbers may be in any base.

14. If $332_{four} + 332_{four} = N_{eight}$, what is the value of N?

    A. 664

    B 332

    C. 174

D. 166

E. 124

15. Write down (in rough) the biggest binary number which can be obtained by using three '0's and three '1's. What is this number expressed as a decimal?

   A. 11,100

   B. 7,000

   C. 700

   D. 112

   E. 56

16. The number 82_{ten} is equivalent to ...

   I. 1010010_{two}

   II. 41_{five}

   III. $6t_{twelve}$

   Which of I, II, III complete the statement correctly?

   A. I, II and III

   B. I and II only

   C. I and III only

   D. II and III only

   E. Neither I, II nor III

17. If $1000_{two} \times 10100_{two} = M_{two}$, what is M?

   A. 101000000

   B. 10100000

   C. 1010000

31

D. 100000

E. 160

18. What is the sum, expressed in base three, of the fractions $\frac{1}{11}$ *three* and $\frac{1}{2}$ *three*?

   A. $\frac{2}{13}$

   B. $\frac{13}{22}$

   C. $\frac{111}{211}$

   D. $\frac{10}{11}$

   E. $\frac{2}{11}$

19. If the total of eight numbers is 520_{six}, what is their average, expressed in base six?

   A 145

   B 104

   C. 65

   D. 40

   E. 24

20. If $\frac{1}{4}$ $_{ten}$ is equivalent to $0\cdot01_{two}$, what binary fraction would $\frac{1}{16}$ $_{ten}$ be equivalent to?

   A. 0·0001

   B. 0·00001

   C. 0·001

   D. 0·0625

   E. 0·000111

Chapter 4 Arithmetic: Rate, Ratio, Percentage

The student should have practice in the wide range of arithmetic problems which involve rates, ratios and percentage.

As in arithmetic conversions (Chapter 2), there are no formulae for this topic. Success comes when the student fully understands the information given in a question and applies mathematical reasoning to this to solve the problem.

Rate
Problems on rates of working, the rate of water filling a tank and similar rates cause difficulty because often the student is not sure where to start.
There are two things he should do:
(i) Write down the information of the question in a form which is useful,
(ii) Find, what we shall call here, a *working unit*.
This is demonstrated in the following example:

'6 men can dig a foundation in 10 days. How long would it take 4 men, working at the same rate, to dig a similar foundation?'
(i) The information of the question can be usefully written as:
   6 men dig 1 foundation in 10 days.
(ii) The *working unit* in this problem is the fraction of the foundation dug by 1 man in 1 day. This can be worked out in two steps:

6 men dig $\frac{1}{10}$ of the foundation in 1 day;

1 man digs $\frac{1}{6 \times 10}$ of the foundation in 1 day.

The problem is finally solved by reasoning as follows:

4 men dig $\frac{4}{6 \times 10}$ of the foundation in 1 day,

i.e. 4 men dig $\frac{1}{15}$ of the foundation in 1 day.
4 men dig 1 foundation in 15 days.

Ratio

Many ratios are expressed in the form $a:b$. For working purposes, treat this as the fraction $\frac{a}{b}$. For example:

'In 1968, the retail price of beer increased in the ratio 6:7. If a bottle of 'Star' now costs 2/11d. retail, what was the retail price before?'

Working in pence, let the previous retail price be x pence per bottle. From the question,

$$6:7 = x:35$$
$$\frac{6}{7} = \frac{x}{35}$$
$$x = \frac{35 \times 6}{7}$$
$$x = 30$$

Thus, the previous retail price of a bottle of 'Star' was 2/6d.

Ratios which connect three or more quantities are expressed in the form $a:b:c$ ... These problems usually involve sharing, as is shown in the following example:

'The Bank of West Africa Ltd. loans £156,000 for a building project. The money is shared between four building companies in the ratio 5:3:3:2. Two of the companies each receive the same amount; how much does each get?'

Here the ratio is expressed as four groups of shares.
The total number of shares = 5 + 3 + 3 + 2 = 13
The *working unit* is 1 share:

$$1 \text{ share} = \frac{£156,000}{13}$$
$$= £12,000$$

The two companies which get the same amount, each get 3 shares:
$$3 \text{ shares} = 3 \times £12,000$$
$$= £36,000$$

Thus, each of the two companies gets £36,000.

Percentage

Remember that $x\%$ is another way of writing $\frac{x}{100}$. Consider the following example:

'Teachers get a 2% rise in their salaries. What is the new salary of a teacher who had been earning £700 per annum?'

$$\begin{aligned}\text{The rise} &= 2\% \text{ of } £700 \\ &= \frac{2}{100} \times £700 \\ &= £14\end{aligned}$$

The teacher now gets £714 per annum.

Other Methods

There are other methods of solving questions in this topic and the student should use the method that he finds easiest.

Good students should be able to solve some of the following questions mentally. This will save time in objective tests.

The above methods are for those students, who, as yet, are unable to get into these problems. Though the worked examples are fairly simple, they demonstrate the method of reasoned argument in solving arithmetic problems.

Class Test: Arithmetic: Rate, Ratio, Percentage
40 minutes 20 questions

1. A farmer dies leaving 3600 acres of land to be shared among his three sons in the ratio 9:7:8. What is the size of the smallest share?

    A. 150 acres

    B. 700 acres

    C. 1050 acres

    D. 1200 acres

    E. 1350 acres

2. 6 is to 108 as $\frac{1}{3}$ is to ...

    A. 3

    B. 6

    C. 9

    D. 18

    E. 54

3. What is 250% of 60?

    A. 24

    B. 90

    C. 96

    D. 120

    E. 150

4. A man saves 40% of his weekly wage. He spends $\frac{2}{3}$ of the remainder on family upkeep and is left with £2 10s to spend on entertainment. How much does he save?

    A. £2 10s.

    B. £4

C. £5

D. £7 10s.

E. £12 10s.

5. It takes three painters 5 days to paint a house. How many days would it take 10 painters to do the job if all the men work at the same rate?

   A. $\frac{1}{2}$ a day

   B. $\frac{2}{3}$ of a day

   C. 1 day

   D. 1$\frac{1}{2}$ days

   E. 16$\frac{2}{3}$ days

6. The switch of a ceiling fan has six 'on' positions. The speed of the fan increases by equal amounts from 100 r.p.m. at position *1* to 400 r.p.m. at position *6*. What is the speed of the fan at position *3*?

   A. 220 r.p.m.

   B. 240 r.p.m.

   C. 250 r.p.m.

   D. 280 r.p.m.

   E. 300 r.p.m.

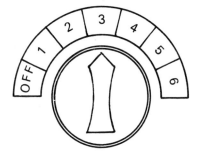

7. The amounts of rainfall for a place for the years 1967 and 1968 were 42" and 49" respectively. By what percentage did the 1968 rainfall exceed the 1967 rainfall?

   A. 16$\frac{2}{3}$%

   B. 14$\frac{2}{7}$%

   C. 14%

   D. 12$\frac{1}{2}$%

   E. 7%

8. In a test, a boy who gets 36 marks is told that he scored 45%. Another boy got 60 marks; what was his percentage?

   A. 60%

   B. 69%

   C. 75%

   D. 79%

   E. 80%

9. 75% of a silver-nickel alloy is nickel. What weight of silver is in a piece of alloy which contains 120 grams of nickel?

   A. 20 gm.

   B. 25 gm.

   C. 30 gm.

   D. 36 gm.

   E. 40 gm.

10. A road which has a gradient of '*1* in *4*' rises 1 foot for every 4 feet of road. Through what height does a car rise when travelling 1 mile along a road whose gradient is *1* in *5*?

    A.   352 ft.

    B.   1056 ft.

    C.   1320 ft.

    D.   5280 ft.

    E.   26400 ft.

11. On a map whose scale is *1*:*n*, two towns are 1 inch apart. If the towns are actually 10 miles apart, what is the value of *n*?

    A.   10

    B.   6336

C. 17600

D. 52800

E. 633600

12. The bobbin of a sewing machine contains enough thread to do 10 yards of stitching. What fraction of the thread will be left on the bobbin if it has already done $4\frac{1}{2}$ feet of stitching?

   A. $\frac{17}{20}$

   B. $\frac{11}{20}$

   C. $\frac{9}{20}$

   D. $\frac{7}{20}$

   E. $\frac{3}{20}$

13. A textbook contains 221 pages with 26 lines per page. If the book were reprinted so that each page had 34 lines, how many pages would it have? (Assume that the lines contain the same number of words in both cases.)

   A. 289

   B. 213

   C. 189

   D. 169

   E. 136

14. A car cost £630 when new. When it was three years old it was sold for £450. What was the percentage loss?

   A. $14\frac{2}{7}\%$

   B. 18%

   C. 20%

   D. $28\frac{4}{7}\%$

   E. 40%

15. A market trader has two boys, aged 11 years and 7 years, who help him. His profit for a day's work was 15s., of which he kept 80%, dividing the remainder between the boys in the ratio of their ages. How much did the younger boy get?

    A. 7d.

    B. 1/2d.

    C. 1/10d.

    D. 3/6d.

    E. 8/2d.

16. The fuel tank of a car holds 8 gallons of petrol which is sufficient for 240 miles of motoring. If the fuel indicator shows that $4\tfrac{1}{2}$ gallons are left in the tank, how far has the car travelled since the tank was full?

    A. 75 mi.

    B. 100 mi.

    C. 105 mi.

    D. 110 mi.

    E. 125 mi.

17. In July 1969, the American spacecraft, *Apollo 11*, splashed down in the Pacific at a speed of 20 m.p.h. On its return from the moon its speed had been 24,000 m.p.h. What percentage of 24,000 is 20?

    A. $\tfrac{1}{1200}$ %

    B. $\tfrac{1}{12}$%

    C.   12%

    D. 1,200%

    E. 120,000%

18. *A* earns 10/- a day, *B* earns £4 10s. a week and *C* earns £8 a fortnight. If there are 6 working days in a week, express these wages in the ratio, *A*'s wage : *B*'s wage : *C*'s wage.

   A. 6 : 9 : 8

   B. 6:9:16

   C. 1:9:8

   D. 1:9:16

   E. 1:6:12

19. In a rest house the cost per person is £2 15s. per day. If the bill for one person who stayed for three days is £9 9s. 9d. find the percentage that has been added as a service charge.

   A. 5%

   B. 9%

   C. 12%

   D. 15%

   E. 20%

20. 3 men can do as much work in one day as 5 boys. A piece of work can be done by 12 men and 10 boys in 6 days. How many days would the work take if 10 boys were employed only?

   A. 2 days

   B. 6 days

   C. 10 days

   D. 12 days

   E. 18 days

Chapter 5 Geometry and Trigonometry in Three Dimensions

Given enough information and working from plane diagrams, the student should be able to calculate the angle between two planes, the angle between a line and a plane, and the angle between two lines in a given solid. He should also be able to calculate the length of a line on or within the solid.

Recognizing Right Angles
Often, in three dimensional problems, the unknown angle (or line) is contained in a right-angled triangle within the solid. Thus, it is essential to be able to recognize any right angles in the solid.

Due to the problems of drawing a solid on plane paper, its right angles are nearly always distorted. When looking at a diagram of a solid, the following should be remembered:

(1) A vertical line meets a horizontal plane at right angles. Consequently, the angle between the vertical line and any line on the horizontal plane drawn from the point where the vertical line meets the plane is a right angle.

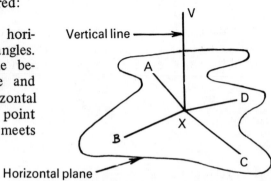

In the diagram,

$V\widehat{X}A = V\widehat{X}B = V\widehat{X}C = V\widehat{X}D = 1$ right angle.

(2) Vertical planes meet horizontal planes at right angles.

In the diagram,
Rectangle ABCD is vertical and rectangle PQRS is horizontal.

$A\widehat{D}P = B\widehat{C}Q$ = 1 right angle.

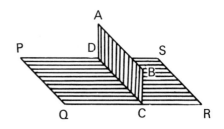

Right-angled Triangles in Solids
Once the right angles have been recognized, the next step is to find a right-angled triangle which contains the required angle (or line).

The following diagrams illustrate (a) the development of a right pyramid to show some of its right-angled triangles, and (b) one of the right-angled triangles which can be developed within a cuboid.

(a) The Right Rectangular Based Pyramid
In a right pyramid, the vertex is directly above the centre of its base.

Its height, VX, can be shown in one of two ways:

X is the point of intersection of the diagonals of the base

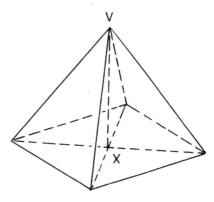

and

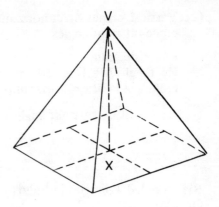

X is the point of intersection of the lines joining the midpoints of opposite sides of the base.

In each of the above diagrams, there are four right-angled triangles standing in vertical planes, each containing the height VX. The triangles have been shaded in the following diagrams:

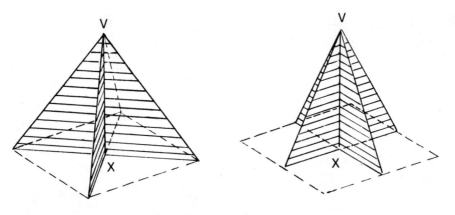

(b) The Cuboid
One of the most useful right-angled triangles within a cuboid is obtained by joining the opposite vertices of the cuboid:

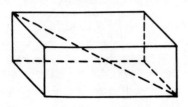 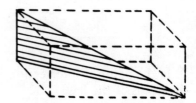

Note that the hypotenuse of this triangle is the longest straight line which can be obtained in the cuboid. How would you calculate its length?

Solving the Right-angled Triangle
When the right-angled triangle which contains the desired angle (or line) has been established, it may be solved by:
(i) the 'angle sum of triangle' theorem,
(ii) Pythagoras' theorem,
(iii) use of the trigonometrical ratios, sine, cosine or tangent of an angle.

Practice
Practice should be gained in breaking down three dimensional figures into triangles which can be dealt with. In addition to right-angled triangles, isosceles triangles are commonly found. (The sloping face of a right pyramid is in the form of an isosceles triangle.)

A common feature of three dimensional work is that often a length is calculated in one triangle so that the calculated length can be used in an adjacent triangle. The following example demonstrates this and some of the other points mentioned above:

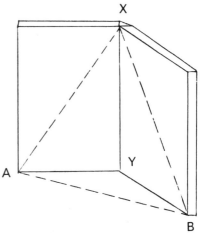

'In the diagram, a book is shown standing open on a horizontal table so that XAB is an equilateral triangle of length of side 10″; If $A\hat{Y}B = 120°$, what is the height, of the book?'

XY is contained in the right-angled triangle XYA. In this triangle, $|AX| = 10″$, so, if the length of AY can be found, the height XY can be calculated using Pythagoras' theorem.

Consider $\triangle AYB$. This triangle is isosceles ($|AY| = |YB|$) so the perpendicular from Y to $\overline{AB}$ meets $\overline{AB}$ at its midpoint, M, and bisects $A\hat{Y}B$. This is shown in the diagram.

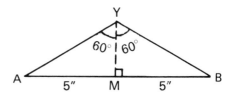

In △AYM,

$$\sin 60° = \frac{5}{|AY|}$$

$$\therefore |AY| = \frac{5}{\sin 60°}"$$

But, $\sin 60° = \frac{\sqrt{3}}{2}$

$$\therefore |AY| = \frac{10}{\sqrt{3}}"$$

Thus, transfering this information to △XYA, the drawing of △XYA is as shown:

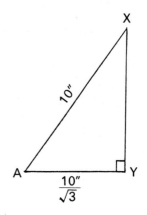

In △XYA,

$$|AX|^2 = |XY|^2 + |AY|^2 \quad \text{(Pythagoras' theorem)}$$

$$100 = |XY|^2 + \frac{100}{3}$$

$$|XY|^2 = 100 - \frac{100}{3}$$

$$= 100 \times \tfrac{2}{3}$$

$$\therefore |XY| = 10\sqrt{\tfrac{2}{3}}"$$

The height of the book is $10\sqrt{\tfrac{2}{3}}$ inches.

Class Test:
Geometry and Trigonometry in Three Dimensions
40 minutes 20 questions

Use the diagram on the right and the given information to answer questions 1, 2, 3.

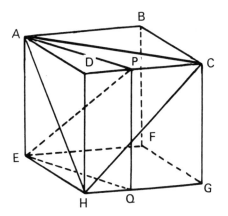

ABCDEFGH is a cube. Face EFGH is horizontal and P and Q are points on the edges CD and GH such that PQ is vertical.

1. What is the name of the angle that plane PQEA makes with the face ADHE?

   A. $P\hat{E}H$

   B. $P\hat{E}D$

   C. $P\hat{A}B$

   D. $P\hat{A}D$

   E. $P\hat{A}H$

2. $P\hat{E}Q$ is the angle between plane EFGH and ...

   A. plane EAPQ

   B. line EP

   C. edge EA

   D. plane EQH

   E. plane EFGQ

3. Consider the lines AC, CH and HA. Which of the following statements does not apply to them?

47

A. One of the three lines lies in a horizontal plane.

B. One of the three lines is horizontal.

C. The three lines form an equilateral triangle.

D. Each line lies in a vertical plane.

E. Two of the three lines are vertical.

4. A 'perfect' orange is spherical and it has 20 equal segments (see the diagram). What is the angle between the semi-circular planes of one of the segments?

A. 18°

B. 20°

C. 30°

D. 36°

E. 40°

Use this diagram and information to answer questions 5 and 6.

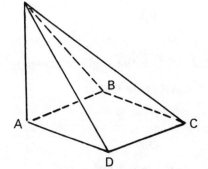

VABCD is a square based pyramid whose vertex, V, is such that VA is perpendicular to ABCD.

5. What is the angle that VC makes with the base ABCD?

A. $V\widehat{C}A$

B. VĈB

C. VĈD

D. CV̂A

E. CV̂D

6. Which of the following is the angle between plane ABCD and plane VCD?

    A. VĈA

    B. VĈB

    C. VD̂A

    D. VD̂C

    E. A right angle

Use this diagram and information to answer questions 7 and 8.

XDCY is a door which is 3 feet by 8 feet. It is shown partly open, leaving XABY as the door frame.

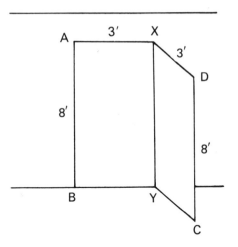

7. If the door has been opened through 132°, what is the size of YB̂C?

    A. 12°

    B. 19°

    C. 24°

    D. 29°

    E. 48°

8. If the door were opened through 180°, what would be the distance of B from D?

   A. 6 ft.
   B. 8 ft.
   C. 10 ft.
   D. 11 ft.
   E. 14 ft.

9.

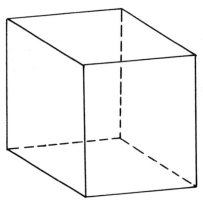

If the centres of each pair of adjacent faces of a cube are joined by straight lines, which one of the following solids would have the joining lines as its edges?

A.

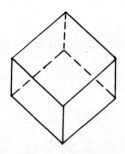

A cube, smaller than the original

B.

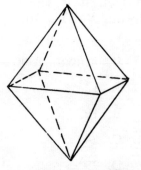

A regular octahedron

C.

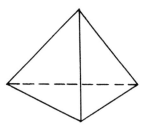

A regular tetrahedron

D.

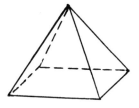

A square based pyramid

E.

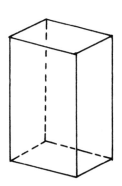

A square based prism

10. ABCD is a rectangular piece of paper, 12" by 8", which has been folded along XY where X and Y are the midpoints of the longer sides. If △XAD is equilateral, what is the distance of A from D?

A. $\sqrt{208}$"
B. 12"
C. $\sqrt{72}$"
D. 10"
E. 8"

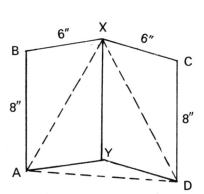

51

11. ABCD is a tetrahedron such that AD is perpendicular to plane DCB. If plane ABC is an equilateral triangle of length of side 2", and if it makes an angle of 60° with plane DCB, what is the length of AD?

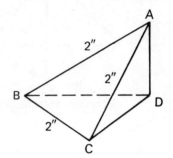

A. $\frac{1}{2}\sqrt{3}$ in.

B. $1\frac{1}{2}$ in.

C. $\sqrt{3}$ in.

D. 2 in.

E. 3 in.

12. XABY is a piece of wire which has been given two right angle bends at A and B so that plane ABX is perpendicular to plane ABY. If the lengths of BY, AB and AX are 3", 4" and 5" respectively, what is the size of $A\hat{Y}X$?

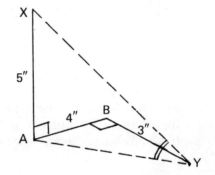

A. 90°

B. 53° 8'

C. 45°

D. 36° 52'

E. 59° 2'

13. PQRSTUVW is a solid formed by truncating a right triangular prism by the plane PQRS. If the edges PU, PS and ST are each 2" long and if edge UT is 4" long, what is the angle between the planes PQVU and SRWT?

A. The planes do not meet, therefore there is no angle between them.

B. 90°

C. 60°

D. 45°

E. 30°

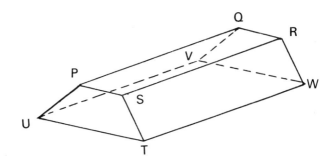

14. PQKLMN is a hipped roof. The ridge PQ, 18 feet long, is placed centrally, 6 feet above the rectangular plane KLMN. If edge ML is 24 feet long, what angle does plane PMN make with plane KLMN?

A. 71° 34'

B. 63° 26'

C. 45°

D. 36° 53'

E. 26° 34'

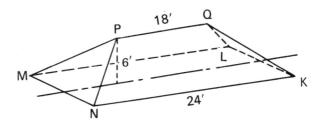

15. In the diagram, a hollow cylinder whose height is the same as its diameter is shown fitting exactly over a solid cone. What is the size of angle x (in the diagram) which is in a plane of symmetry of the pair of objects?

A. 22° 30'

B. 26° 34'

C. 30°

D. 45°

E. 60°

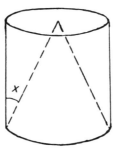

16. VABCD is a rectangular based pyramid, with ABCD as base. The sloping edges are each 17 cm. long, edge AB is 16 cm. long and edge BC is 24 cm. long. What is the height of the vertex, V, above the base?

A.   17 cm.
B.   15 cm.
C.   √145 cm.
D.   12 cm.
E.   9 cm.

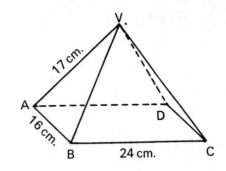

17. In a football match, a player takes a penalty and (unfortunately) the ball hits the top of the upright where it joins the crossbar. The penalty spot, which is situated centrally, is 12 yards from the goal-line and the uprights are 8 feet high and 8 yards apart. How far did the ball travel?

A.   12 yd.
B.   √167·1 yd.
C.   √217·1 yd.
D.   √224 yd.
E.   18⅔ yd.

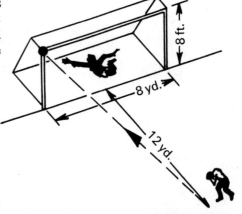

Use this diagram and information to answer questions 18 and 19.

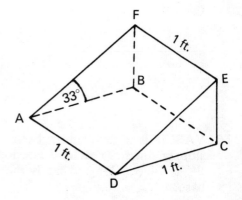

ABCDEF is a wedge. Planes ABCD and BCEF are horizontal and vertical respectively. ABCD is a square of side 1 foot and FÂB = 33°.

18. What is the tangent of F$\hat{A}$E?

   A. cos 33°

   B. tan 33°

   C. $\dfrac{1}{\tan 33°}$

   D. $\dfrac{1}{\cos 33°}$

   E. $\dfrac{\tan 33°}{\sqrt{2}}$

19. What is the tangent of E$\hat{A}$C?

   A. cos 33°

   B. tan 33°

   C. $\dfrac{1}{\tan 33°}$

   D. $\dfrac{1}{\cos 33°}$

   E. $\dfrac{\tan 33°}{\sqrt{2}}$

20. A hut is in the form of a right cone of height 8 feet placed on top of a cylinder of height 10 feet. The diameter of the cone and the cylinder is 12 feet. If the elevation of the sun is 45°, what is the area of the shadow of the hut on the ground?

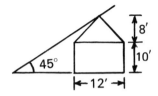
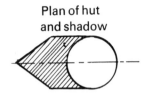

Plan of hut and shadow

   A. 168 sq. ft.
   B. 216 sq. ft.
   C. (168 − 18π) sq. ft.
   D. (216 − 18π) sq. ft.
   E. 456π sq. ft.

55

Chapter 6 Graphs: Interpretative and Algebraic

Interpretative Graphs
The student should be able to read and interpret the information displayed on a wide variety of descriptive and statistical graphs. Such graphs fall into two main categories:

1. Information Represented by a Continuous Line
Many interpretative graphs show how one event varies with another by means of a continuous line drawn within rectangular axes. This is convenient for showing how certain events vary with time (always represented on the horizontal axis). For example, graphs can be drawn to show:
(a) The distance of a car from a town at any time after its departure.
(b) The height of a ball above the ground at any time during its flight.
(c) The temperature of a furnace at any time of day.

Other graphs, however, are not connected with time. For example, graphs can be drawn to show:
(a) How industrial profit varies with production output.
(b) How the petrol consumption of a car varies with its speed.

2. Information Represented by Area
Statistical data is often illustrated by graphs in which areas are taken to represent groups of numbers. Two common examples of this are:

(a) *Bar Charts*
   A bar chart to show the inches of rainfall for each month in a certain year.
   The height of each column is proportional to the rainfall for its corresponding month.

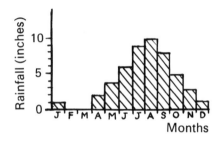

(b) *Pie Charts*
   A pie chart showing the proportions of population in the 5 states of a certain country.

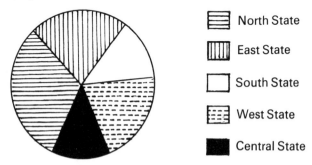

The area of the circle represents the total population of the country, thus the population of each state is proportional to the angle of the sector which represents that state.

Algebraic Graphs

For Certificate purposes, the student, given $y = f(x)$ (i.e. y is a function of x, such as $2x + 3$; $x^3 - 5$; $1/x$), should be able to plot corresponding values of x and y within axes to form either a straight line or a curve. He should be able to solve simultaneous equations and quadratic equations from a given graph and should be able to calculate the gradient of a curve at any point from a tangent construction.

It is not often realized that the actual function of x provides a clue to the general shape of its graph. The following *sketch graphs* should be noted:

Linear functions

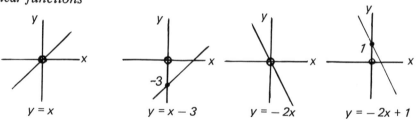

$y = x$ $y = x - 3$ $y = -2x$ $y = -2x + 1$

Note (a) the coefficient of x gives the gradient of the line;
(b) the constant term gives the intercept on the y-axis.

Quadratic functions

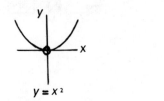
$y = x^2$

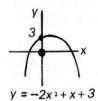
$y = -2x^2 + x + 3$

Note (a) the curve is in the shape of a parabola with a vertical axis of symmetry;
(b) if the coefficient of x^2 is positive the parabola is upright;
(c) if the coefficient of x^2 is negative the parabola is inverted.

Cubic functions

(i)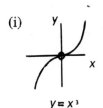
$y = x^3$

(ii)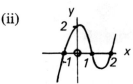
$y = (x+1)(x-1)(x-2)$

(iii)
$y = (x+1)(x-1)(2-x)$

Note the effect on the shape of the curve when the coefficient of x^3 is positive (i and ii) and when the coefficient of x^3 is negative (iii).

Inverse functions

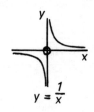
$y = \frac{1}{x}$

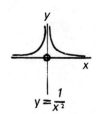
$y = \frac{1}{x^2}$

Note (a) the curve $y = 1/x$ appears in the 1st and 3rd quadrants and is symmetrical about the lines $y = \pm x$;
(b) the curve $y = 1/x^2$ appears in the 1st and 2nd quadrants and is symmetrical about the y-axis;
(c) both curves are undefined when $x = 0$.

Reading Graphs

When reading graphs, attention should be given to the following:
(1) As you work, check the scale of each axis. Very often, different scales are used on the two axes.
(2) On 1 inch graph paper it is quite possible to read to 0·01 of an inch, therefore do not round off any values but estimate them as accurately as you can.
(3) As you work, check whether the values and gradients that are found are positive or negative.

Class Test: Graphs: Interpretative and Algebraic
40 minutes 20 questions

The graph shows the journeys of a lorry and a car from *Minna* to *Bida* and from *Bida* to *Minna* respectively. Each journey was via *Zungeru* and the journeys were made on the same day. Use the graph to answer questions 1, 2, 3, 4, 5.

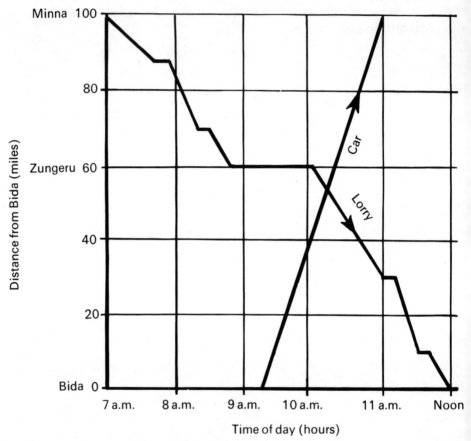

1. How many stops did the lorry make *between Minna* and *Bida*?

    A. 1

    B. 4

    C. 5

    D. 6

    E. 7

2. What was the average speed of the car for its journey from *Bida* to *Minna*?

   A. 80 m.p.h.

   B. 50 m.p.h.

   C. 100 m.p.h.

   D. 60 m.p.h.

   E. 444 m.p.h.

3. What was the average speed of the lorry for its journey from *Minna* to *Bida*?

   A. $33\frac{1}{3}$ m.p.h.

   B. 30 m.p.h.

   C. 25 m.p.h.

   D. 20 m.p.h.

   E. $16\frac{2}{3}$ m.p.h.

4. How long did the lorry stop at *Zungeru*?

   A. 1 hr. 20 min.

   B. 1 hr. 12 min.

   C. 1 hr. 10 min.

   D. 1 hr. 6 min.

   E. 1 hr. 2 min.

5. How far were the vehicles from *Zungeru* when they passed each other on the road?

   A. 54 mi.

   B. 50 mi.

   C. 46 mi.

   D. 6 mi.

   E. 3 mi.

The graph shows the amounts of income tax to be paid annually on the incomes of:
(S) : single men with no allowances
(C) : men who have two children receiving full-time education.
Use the graph to answer questions 6, 7, 8, 9.

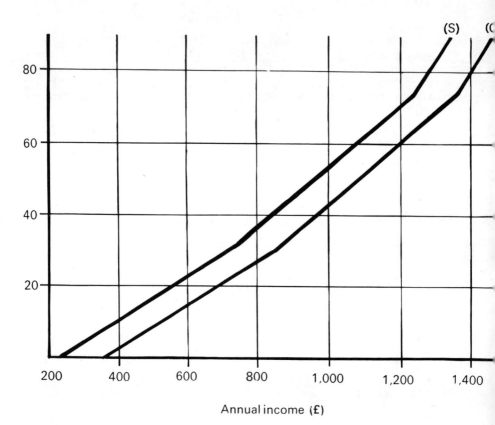

6. How much can a single man earn without paying any tax?

   A.   £0

   B.   £40

   C.   £200

   D.   £220

   E.   £240

7. How much tax does a single man pay if he earns £1,100 in a year?

   A. £63

   B. £72

   C. £60

   D. £54

   E. £52

8. A single man and a man with two sons at the Secondary School both get an income of £1,000 per annum. What is the difference in the tax paid by the two men?

   A. £120

   B. £54

   C. £44

   D. £10

   E. They both pay the same amount of tax.

9. If two men, one from each of the categories (S) and (C), each pay the same amount of tax in a year, which of the following statements is (are) true?

   I. Their incomes must be the same.

   II. The man in category (C) has a higher income than the man in category (S).

   III. Their incomes differ by £120.

   A. None of the statements is correct.

   B. I only

   C. II only

   D. III only

   E. Both II and III

10. The accompanying graph shows how the shade temperature varies with time during a day in the Harmattan season. Which of the following statements is not true?

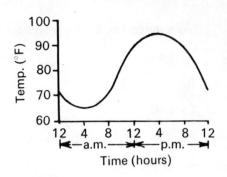

   A. The difference between the maximum and minimum temperatures was 30° F.

   B. At 6 a.m. the temperature was rising.

   C. At 6 p.m. the temperature was falling.

   D. The maximum temperature occurred at 12 noon.

   E. The minimum temperature occurred at 4 a.m.

Use the following graph to answer questions 11 and 12.

A Pie Chart to show the proportions of workers in different jobs in a Ministry of Works yard which employs a total of 240 men.

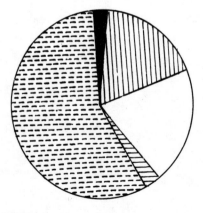
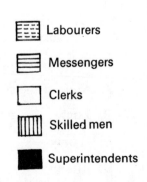

11. If the yard employs 138 labourers, what is the angle of the sector which represents labourers?

    A. 207°

    B. 276°

    C. 138°

    D. 222°

    E. 92°

12. If the angle of the sector which represents clerks is 72°, how many clerks are employed by the yard?

    A. 36

    B. 48

    C. 60

    D. 72

    E. 96

13. The bar chart shows the numbers of boys in a class whose hometowns are between 0 and 50 mi., 50 and 100 mi., ... 200 and 250 mi. from the school. How many boys are in the class?

    A. 29

    B. 28

    C. 24

    D. 27

    E. 30

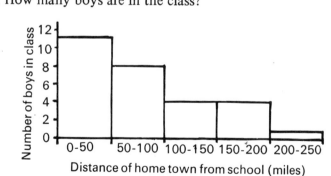

14. The diagram shows part of the graph of $y = x^2 - 8x$. S lies on the axis of symmetry of the curve. What is the value of y at S?

    A. -16

    B. -8

C. −4
D. −2
E. 0

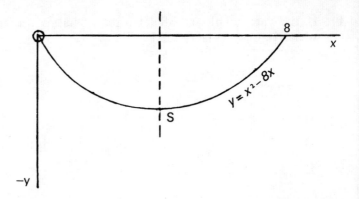

15. A class was asked to draw sketch graphs of I, $y = x^2$, II, $y = 1/x^2$, III, $y = 1/x$. Here is what one boy drew:

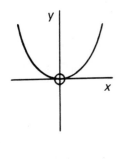

I

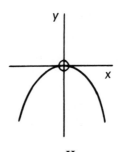

II

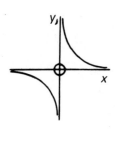

III

Which of these sketches is (are) correct?

A. I only
B. Both I and II
C. Both I and III
D. Both II and III
E. I, II and III are all correct.

The curve, $y = 5 + 2x - 2x^2$, and the line, $y = -2x - 4$, have been drawn on the same axes.
Use the graph to answer questions 16, 17, 18, 19, 20.

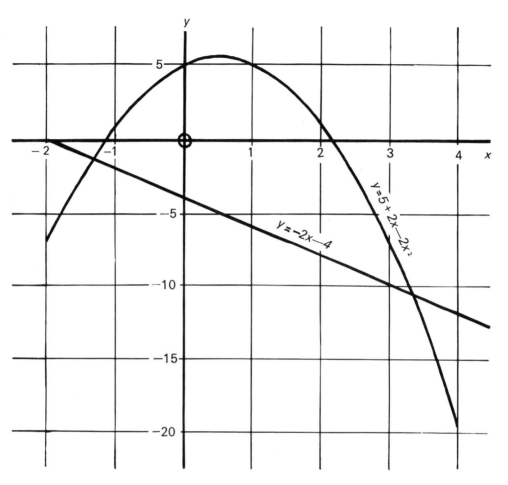

16. For what value of x is $5 + 2x - 2x^2$ a maximum?

    A. +6
    B. +5·5
    C. +4
    D. +1
    E. +½

17. What are the solutions of the equation, $2x^2 - 2x - 5 = 0$?

    A. + 1·16 and + 2·16
    B. + 1·16 and − 2·16

C. −1·16 and +2·16

D. −2·84 and +2·16

E. Impossible, as the curve $y = 2x^2 - 2x - 5$ has not been given.

18. By drawing the tangent to the curve at $x = -0.5$, or otherwise, deduce the gradient of the curve at the point where $x = -0.5$.

    A. −5

    B. −4

    C. +0·4

    D. +0·8

    E. +4

19. Use the graph to solve the equation $-2x - 4 = 5 + 2x - 2x^2$. $x =$

    A. −1·35 and +3·35

    B. −1·3 and −10·7

    C. +1·35 and +3·35

    D. −2 and +4

    E. −2·65 and +3·35

20. What are the solutions of the equation $1 + 2x - 2x^2 = 0$?

    A. 0 and +1

    B. −0·37 and +1·37

    C. −1·67 and +2·67

    D. −1 and +2

    E. −1·9 and 2·9

Chapter 7 Latitude and Longitude

The student should know the definitions of (i) angle of latitude, (ii) angle of longitude, (iii) the nautical mile. Using this knowledge and his knowledge of circles, the student should be able to calculate distances between two points on (a) Great Circles, (b) Small Circles.

As in the three dimensional geometry of Chapter 5, much of the difficulty in this topic arises out of the problem of representing a three dimensional object, in this case a sphere, on plane paper. Once again, the solid must be broken down into units which can be dealt with. Therefore, the globe of the earth must be broken down into circles.

The Length of a Circular Arc

The length, l, of an arc of a circle, radius r, which subtends an angle $\theta°$ at the centre of the circle, is given by:

$$l = \frac{\theta}{360} 2\pi r$$

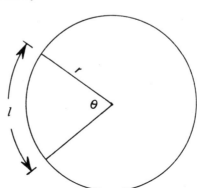

The following example shows how this knowledge can be used in latitude and longitude problems:

'A plane leaves Lagos, 4°E, 6°N, and flies due north to Algiers, 4°E, 36°N. It then leaves Algiers and flies due west at 600 m.p.h.
(a) What distance does it fly between Lagos and Algiers?
(b) What is the position of the plane 4 hours after taking off from Algiers? Assume that the earth is a sphere of radius 3960 statute miles and that $\log \pi = 0.4971$.'

First draw a *good* sketch of the journey.

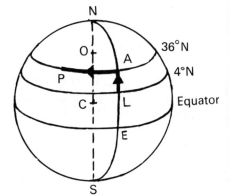

In the diagram, C is the centre of the earth and O is the centre of the parallel of latitude 36°N. L and A denote the positions of Lagos and Algiers and P is the position of the plane 4 hours after leaving Algiers. NS is the polar axis.

(a) The arc LA is on the great circle NALES, centre C. Extracting this circle from the sketch,

$A\hat{C}L = 36° - 6°$
$\therefore A\hat{C}L = 30°$

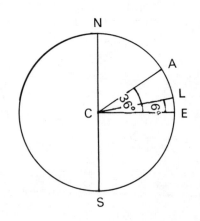

Great circle
Radius = 3960 st.mi.

$$|\text{Arc LA}| = \frac{30}{360} 2\pi \; 3960 \text{ st.mi.}$$

No.	Log.
660	2·8195
π	0·4971
2073	3·3166

After simplifying, |Arc LA| = 660π st. mi.
From tables, |Arc LA| = 2073 st. mi.

The plane flies 2073 statute miles between Lagos and Algiers.

(b) Distance travelled by the plane in 4 hours = 4 x 600 st.mi
= 2400 st.mi
Thus, |Arc AP| = 2400 st.mi

Arc AP is on the small circle of latitude, 36°N, centre O. Extracting this circle from the sketch,

Let $P\hat{O}A$ be θ.

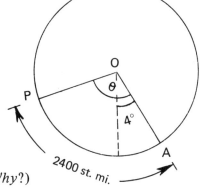

Circle of latitude, 36°N.
Radius = 3960 Cos 36° st.mi. (*Why?*)

$$2400 = \frac{\theta}{360} 2\pi \; 3960 \text{ Cos } 36°$$

After simplifying, $\theta = \frac{2400}{22\pi \text{ Cos } 36°}$ degrees

No.	Log.
2400	3·3802
22	1·3424
π	0·4971
cos 36°	1̄·9080
	1·7475
42·92	1·6327

From tables, $\theta = 42.92°$

To nearest minute, $\theta = 42° \; 55'$

The plane has flown westwards through an angle of 42° 55′ from a position 4°E. Thus, its new longitude is (42° 55′−4°) W., or 38°55′W. The position of the plane four hours after taking off from Algiers is 36°N. 38°55′W.

71

Method
(1) From a good sketch of the Earth, extract the circle which contains the arc, or angle, in which you are interested.
(2) Insert any given information on this circle. The information should include *two* of the following; the third is to be calculated:
   (a) the radius of the circle,
   (b) the angle which the arc subtends at the centre of the circle,
   (c) the length of an arc.
(3) Apply your knowledge of the length of the arc of a circle to solve the problem.

Nautical Mile
1 nautical mile is the length of 1 minute of arc of a great circle; i.e. the distance travelled along a line of longitude between two points whose latitudes differ by 1 minute.

The length of an arc of a Great Circle may be easily calculated in nautical miles:

If the angle subtended by the arc at the Earth's centre is $\phi°$, then the length of the arc is 60ϕ nautical miles.

Class Test: Latitude and Longitude
40 minutes 20 questions

Note (1) The aim of this section is to improve the student's ability in recognizing and interpreting information on latitude and longitude. Thus there is no lengthy calculation which is common in written problems. However, in some examples, some easy arithmetic simplification must be done before the student makes his choice.

(2) Unless otherwise stated, assume that the radius of the Earth is 3960 statute miles.

Use the following diagram and information to answer questions 1-5.

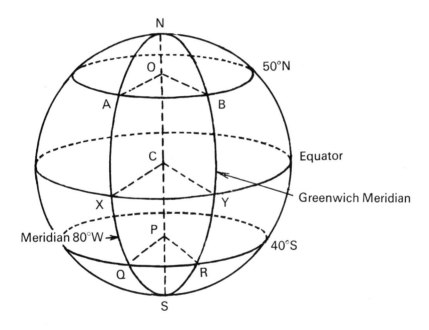

1. What is the size of AÔB?

    A. 30°

    B. 40°

    C. 50°

    D. 80°

    E. 120°

2. Which of the following angles is equal to 40°?

   A. $X\hat{C}Q$

   B. $C\hat{X}Q$

   C. $P\hat{C}Q$

   D. $C\hat{Q}X$

   E. $X\hat{C}P$

3. What is the size of $B\hat{C}R$?

   A. 170°

   B. 90°

   C. 80°

   D. 50°

   E. 40°

4. What angle does the arc QS subtend at C (the centre of the Earth)?

   A. 30°

   B. 40°

   C. 50°

   D. 80°

   E. 90°

5. What line gives the distance of the plane of the parallel of latitude 50°N. from the plane of the parallel of latitude 40°S.?

   A. $\overline{OP}$

   B. Arc AQ

   C. Arc BR

   D. Arc AR

   E. Arc BQ

6. An arc on the equator subtends an angle of 14° at the centre of the Earth. What is the length of the arc in nautical miles?

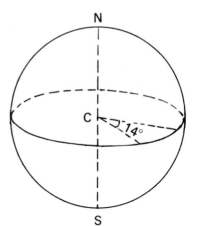

A. $\frac{14}{360} 2\pi$ 3960 naut. mi.

B. 3960 cos 14° naut. mi.

C. 14 naut. mi.

D. 50400 naut. mi.

E. 840 naut. mi.

7. A point A is on the Greenwich meridian in latitude 8°N. A point B is 750 nautical miles due north of A. What is the latitude of B?

A. 4° 30′ S.

B. 12° 30′ N.

C. 20° 30′ N.

D. 75° N.

E. 83° N.

8. X is a point on the parallel of latitude ϕ°N., centre O, and Y is a point on the Equator, centre C.

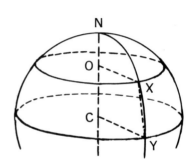

If the radius of the Earth is R miles in length, which of the following statements if not true if X lies due north of Y?

A. $|OX| = R \cos \phi°$ mi.

B. $|OX| = R \sin (90 - \phi)°$ mi.

C. $|OC| = R \sin \phi°$ mi.

D. $C\widehat{O}X = 1$ right angle

E. $O\widehat{X}Y = 1$ right angle

Use the following diagram and information to answer questions 9-12.

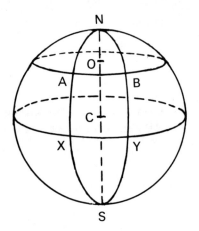

In the diagram, NS is the polar axis and the circle through X and Y, centre C, is the equator. Also shown are two meridian half circles and a parallel of latitude.

9. If the parallel of latitude is 44° N., what is the length of the arc BY?

A. $\frac{44}{360} 2\pi$ 3960 st.mi.

B. $\frac{46}{360} 2\pi$ 3960 st.mi.

C. $\frac{44}{360} 2\pi$ 3960 cos 44° st.mi.

D. $\frac{44}{360}$ 3960 st.mi.

E. 2π 3960 cos 44° st.mi.

10. If the latitude is 60°N. and the longitudes differ by 150°, which of the following is the distance between A and B along the parallel of latitude?

A. 1650π st.mi.

B. 2310π st.mi.

C. 3300π st.mi.

D. 4620π st.mi.

E. 330π st.mi.

11. If the latitude is 37°N., which of the following is the length of the Arc BN?

   A. $\frac{37}{360}$ 2π 3960 st.mi.

   B. $\frac{63}{360}$ 2π 3960 st.mi.

   C. $\frac{53}{360}$ 2π 3960 st.mi.

   D. $\frac{53}{360}$ 2π 3960 cos 37° st.mi.

   E. More information is needed.

12. If the latitude is 42°N., which of the following is |NO|?

   A. 3960 cos 42° st.mi.

   B. 3960 (1 − cos 42°) st.mi.

   C. 3960 sin 42° st.mi.

   D. 3960 (1 − sin 42°) st.mi.

   E. ½ of 3960 st.mi.

13. An aircraft flies from P (45°N., 50°W.) to Q (20°S., 50°W.). It then flies from Q to R (20°S., 30°E.). Which of the following diagrams gives the route (shown by the heavy arrowed lines) of the aircraft?

A.
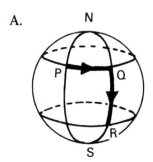

B.
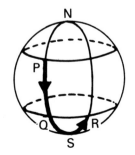

C.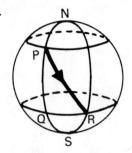

D.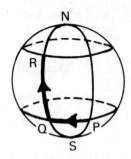

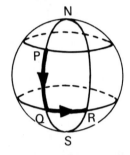

14. What is the distance along the meridian from Alexandria (32°N., 30°E.) to Ladysmith (28°S., 30°E.)?

    A. 88π st.mi.

    B. 1100π st.mi.

    C. 1320π st.mi.

    D. $88\pi \cos 30°$ st.mi.

    E. $1320\pi \cos 30°$ st.mi.

15. What is the distance along the parallel of latitude between Timbuktu (18°N., 3°W.) and Berber (18°N., 34°E.)?

    A. $792\pi \cos 3°$ st.mi.

    B. $792\pi \cos 34°$ st.mi.

    C. $407\pi \cos 18°$ st.mi.

    D. $671\pi \cos 18°$ st.mi.

    E. $814\pi \cos 18°$ st.mi.

16. Two places are in the same latitude north of the Equator and their longitudes differ by 180°. Aeroplane A flies between the places along the parallel of latitude; aeroplane B makes the journey along the great circle route via the North Pole. Which aeroplane travels the shorter distance?

    A. Aeroplane A

    B. Aeroplane B

    C. Each aeroplane travels the same distance.

    D. We cannot say, unless we know the actual latitude.

    E. It depends what the speeds of the aeroplanes are.

17. A seaplane takes off from one of the Cook Islands, 156°W., 21°S., and flies non-stop to Lake Ngami, 24°E., 21°S., on a great circle route via the South Pole. What distance did the seaplane travel?

    A. $\frac{138}{360} 2\pi$ 3960 st.mi.

    B. $\frac{69}{360} 2\pi$ 3960 st.mi.

    C. $\frac{42}{360} 2\pi$ 3960 st.mi.

    D. $\frac{21}{360} 2\pi$ 3960 st.mi.

    E. $\frac{1}{2}$ of 2π 3960 st.mi.

18. A piece of string is stretched round the Equator of a classroom globe of the Earth whose radius is 6"; What is the length of the string?

    A. 2π 3960 mi.

    B. π ft.

    C. $\frac{1}{2}\pi$ ft.

    D. $\frac{1}{4}\pi$ ft.

    E. 1 ft.

19. The length of a parallel of latitude, $\theta°$N., is X statute miles. If the radius of the Earth is R statute miles, what is θ in terms of X, π, and R?

A. $\cos\theta = \dfrac{2\pi R}{X}$

B. $\cos\theta = \dfrac{X}{2\pi R}$

C. $\cos\theta = \dfrac{X}{R}$

D. $\cos\theta = \dfrac{2\pi X}{R}$

E. $\cos\theta = \dfrac{R}{2\pi X}$

20. If the radius of the parallel of latitude $\phi°$N. is half the length of the radius of the equator, what is the value of ϕ?

    A. 30

    B. 45

    C. 60

    D. 75

    E. More information is needed.

Chapter 8 Linear Inequalities and Undefined Fractions

Linear Inequalities

The student should be sure of the meaning of the symbols $>$ and $<$ and be able to find a range of values of x which satisfies a given linear inequality in x.

A linear function of x is one in which x is raised to the power 1 only. Thus, $3x - 2$, $(x/2) + 1$, $4 - 7x$ are examples of linear functions of x. The word *linear* results from the fact that the graph of such a function is a straight *line*.

With one very important exception, the methods used in solving algebraic equations can be used in obtaining ranges of values from given inequalities. The exception is that while it is possible to multiply or divide both sides of an *equation* by a *negative number* without changing the truth of the equation, *the truth of a given inequality is reversed* if each side is multiplied or divided by a negative number. Two simple examples demonstrate this:

(1) It is true to say that -24 is a smaller number than -6,
    i.e. $-24 < -6$ (true)

Consider what happens when both sides of the inequality are divided by -6.

Is $\dfrac{-24}{-6} < \dfrac{-6}{-6}$?

i.e., is $4 < 1$?

Of course, 4 is not a smaller number than 1 and the statement is untrue.

(2) Multiplication by a negative number gives a similar reversal of the truth.

e.g. $7 > \frac{3}{4}$ (true)
If both sides of the inequality are multiplied by -4
Is $7 \times (-4) > \frac{3}{4} \times (-4)$?
i.e., is $-28 > -3$?
-28 is not greater than -3, thus the final statement is untrue.

Reading Inequalities

The statement, $4 < x$, can *either* be read as $4 < x$ (4 is less than x) *or* as $x > 4$ (x is greater than 4). This is useful to remember, for, in dealing with inequality problems, x is just as likely to appear on one side of the inequality as the other.

Examples

The following examples show the general algebraic method for finding the range of values of x which satisfies a given inequality in x.

(1) 'Find the range of values of x for which $3x - 2 < 5x - 8$.'
$$3x - 2 < 5x - 8$$
Adding 8 to, and subtracting $3x$ from, both sides,
$$8 - 2 < 5x - 3x$$
i.e. $\qquad 6 < 2x$
Dividing both sides by 2,
$$3 < x$$
Thus, the inequality is satisfied if $x > 3$.

(2) 'For what range of values of x is the inequality, $3(5x - 12) > 9(x - 4)$, satisfied?'
$$3(5x - 12) > 9(x - 4)$$
Multiplying out the brackets,
$$15x - 36 > 9x - 36$$
Adding 36 to, and subtracting $9x$ from, both sides,
$$15x - 9x > 0$$
i.e. $\qquad 6x > 0$
Dividing both sides by 6,
$$x > 0$$
Thus, if $x > 0$, the inequality is satisfied.

Undefined Fractions

Division by zero is impossible. Thus if the denominator of an algebraic fraction has the value zero for some value of x the fraction is said to be undefined for that value of x.

The word *undefined* is used because the value of a fraction which has a denominator of zero is *not finite* (i.e., the value is unlimited). This can be demonstrated on the graph of $y = 1/(x - 2)$:

From the graph, it can be seen that as the value of x approaches 2, the value of y is *either* an uncountable low number (as the value of x increases towards 2) *or* an uncountable high number (as the value of x decreases towards 2). Due to the uncertainty of the value of y when $x = 2$ we say that the fraction is undefined when $x = 2$.

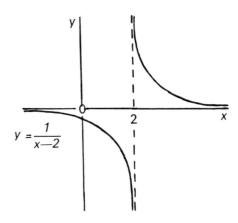

Examples
The following examples show that for certain values of x, the denominators of the given algebraic fractions become zero. The fractions are undefined for those values of x.
  'For what values of x are the following fractions undefined?

$$\frac{x}{x+2}; \quad \frac{x+1}{2x-3}; \quad \frac{1}{7(5x-1)}; \quad \frac{3x+1}{x^2+2x}; \quad \frac{1}{x^2-x-20},$$

$\dfrac{x}{x+2}$ is undefined if $x + 2 = 0$
  i.e., if $x = -2$

$\dfrac{x+1}{2x-3}$ is undefined if $2x - 3 = 0$
  i.e., if $2x = 3$
  i.e., if $x = \dfrac{3}{2}$

$\dfrac{1}{7(5x-1)}$ is undefined if $7(5x - 1) = 0$
  i.e., if $5x - 1 = 0$ (since $7 \neq 0$)
  i.e., if $x = \dfrac{1}{5}$

$\dfrac{3x+1}{x^2+2x}$ is undefined if $x^2 + 2x = 0$
  i.e., if $x(x + 2) = 0$
  i.e., *either* if $x = 0$
    or if $x = -2$

$\dfrac{1}{x^2 - x - 20}$ is undefined if $x^2 - x - 20 = 0$

i.e., if $(x - 5)(x + 4) = 0$
i.e., *either* if $x = 5$
  or if $x = -4$

Class Test: Linear Inequalities and Undefined Fractions
40 minutes 20 questions

1. What is the range of values of x for which $x - 7 > 0$?

    A. $x > -7$

    B. $x > 7$

    C. $x < 7$

    D. $x < -7$

    E. $x > 0$

2. What is the range of values of y if $14y + 42 < 0$?

    A. $y < -42$

    B. $y < 3$

    C. $y < -3$

    D. $y > 3$

    E. $y > -3$

3. Which of the inequalities below is equivalent to $x < 2$?

    A. $2 - x < 0$

    B. $x + 2 < 0$

    C. $x + 2 > 0$

    D. $x - 2 > 0$

    E. $x - 2 < 0$

4. What is the range of values of x for which $x + 8 > 10$?

    A. $x > 2$

    B. $x > -2$

    C. $x > 18$

D. $x > -18$

E. $x > 1\frac{1}{4}$

5. What is the range of values of x for which $14 > 6 + 4x$?

    A. $x > 8$

    B. $x > 2$

    C. $x < 8$

    D. $x < 2$

    E. $x < 5$

6. What is the range of values of x for which $2x + 6 > 2(5 + x)$?

    A. The given inequality is not true for any value of x

    B. $x > 1$

    C. $x > 4$

    D. $x < -1$

    E. $x < -4$

7. What is the range of values of x for which $8x + 42 < 3 - 5x$?

    A. $x < -15$

    B. $x < 13$

    C. $x < -13$

    D. $x < -3$

    E. $x < 3$

8. What is the range of values of y for which $3y + 10 < 4y + 11$?

    A. $y > 1$

    B. $y > -1$

    C. $y < 1$

    D. $y < -1$

E. $y > 7$

9. If $3(x - 4) > 2(4 - x)$, what is the range of values of x?

   A. $x > -4$

   B. $x > 4$

   C. $x > 20$

   D. $x > -5$

   E. $x > 5$

10. What is the range of values of x for which $\frac{x}{3} + \frac{x}{2} > 3$?

    A. $x > \frac{2}{3}$

    B. $x > 1\frac{1}{3}$

    C. $x > 1\frac{1}{2}$

    D. $x > 3\frac{2}{3}$

    E. $x > 9$

11. If $3(x - 2) - \frac{x}{3} + 6 > 0$, what is the range of values of x?

    A. $x < 0$

    B. $x > 0$

    C. $x > 4\frac{1}{2}$

    D. $x > 3$

    E. $x > \frac{1}{8}$

12. At the beginning of term, a school has 19 classes with an average of x boys in each class. The school's 400 desks are sufficient to seat the boys. Later, 24 boys are transferred to the school and the principal finds that he has to buy more desks. Which of the following statements represents this information?

    A. $19x + 24 = 400$

    B. $19x + 24 = 423$

C. $19x + 24 > 400$

D. $19x + 24 < 400$

E. $\quad\quad 19x > 424$

13. What is the range of values of x of any point which lies within the shaded area of the given graph?

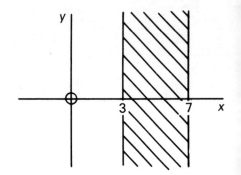

A. $x > 3$

B. $x < 7$

C. $3 < x < 7$

D. $3 > x > 7$

E. $x > 0$

14. For what value of y is the fraction $\dfrac{1}{3(y-5)}$ not defined?

A. -15

B. $\quad 15$

C. $\quad -5$

D. $\quad 5$

E. $\quad 3$

15. One of the following fractions is not defined if $x = 3$. Which one?

A. $\dfrac{1}{3x - 1}$

B. $\dfrac{1}{x + 3}$

C. $\dfrac{1}{x - 3}$

D. $\dfrac{1}{3(x + 1)}$

E. $\dfrac{1}{3(x-1)}$

16. For what value(s) of x is the expression $\dfrac{x}{2x+1} + \dfrac{1}{x-2}$ not defined?

    A. 2 only
    B. ½ only
    C. Both 2 and ½
    D. Both 2 and -1
    E. Both 2 and $-½$

17. For what values of x is the fraction $\dfrac{7}{x^2 - 5x - 24}$ not defined?

    A. $+3$ and -8
    B. $+3$ and $+8$
    C. -3 and -8
    D. -3 and $+8$
    E. $+5$ and 0

18. For what value(s) of x is the fraction $\dfrac{x-1}{x^2 - 7x}$ undefined?

    A. 0 only
    B. 1 only
    C. 7 only
    D. 0, 1 and 7
    E. 0 and 7, only

19. Which of the fractions, I, II, III, IV is undefined if $x = \pm 6$?

    I $\dfrac{1}{1+6x}$ II $\dfrac{1}{1-6x}$ III $\dfrac{1}{x^2 - 36}$ IV $\dfrac{1}{(x+6)(x-6)}$

89

A. I and II only

B. II and III only

C. III and IV only

D. I, II, III, and IV

E. IV only

20. For what values of x is the expression, $\dfrac{1}{x^2 - 10x + 25} + \dfrac{2x}{x^2 + 7x + 10}$, not defined?

A. $-2, 2, -5$ and 5

B. $-2, -5$ and 5

C. $-2, 2$ and 5

D. -2, and 5 only

E. -2 and -5 only

Chapter 9 Plan and Elevation

The student should be familiar with what the terms *plan* and *elevation* mean. Further, it will help him if he has *seen and drawn* some of the simpler solids, such as right prisms (cube, cuboid, triangular), right pyramids (cone, square based, triangular based) and the hemisphere.

Skill in drawing plans and elevations can be increased if the student can develop the ability to draw accurately what is visible from a particular line of vision, *not* what he *imagines* is there. To this end, he should, whenever possible, work with models of the solids that he is drawing. The following is a list of some of the simple solids which can easily be obtained:

Cube—cube sugar.
Cuboid—matchbox, chalkbox, shoebox.
Prisms—'cut-offs' from the Woodwork shop, glass prisms from the Physics laboratory.
Cone—the sharpened end of a round pencil, the roof of a round hut is a good approximation to a cone.
Pyramid—classroom models are difficult to obtain unless made. Examples to be seen outside are the roofs of some small buildings and the groundnut pyramids in the north of Nigeria.

Some General Rules

Plan
(a) Any edges which are horizontal on the solid will appear full length on the finished plan.

(b) Any edges which are sloping on the solid (i.e. edges which are at some angle to the horizontal) will appear shortened on the plan.
(c) Any vertical edges on the solid will appear as points on the plan.

Elevation
(a) Any edges which are vertical on the solid will appear full length on the finished elevation.
(b) Any edges of the solid which are *in the vertical plane of the elevation,* or are *in planes parallel to that plane,* will appear full length on the elevation.
(c) Edges which make an angle with the plane of the elevation will appear shortened on the elevation.
(d) Any horizontal edges which are perpendicular to the plane of the elevation (i.e. parallel to the line of vision of the draughtsman) will appear as points on the elevation.

The following example will demonstrate the above rules:

'A square-based pyramid is such that its vertex, E, is directly above the point A of its base ABCD. The edges EA and AB are each 2" in length. The solid is truncated by plane BXYC, where X and Y are the mid-points of edges EA and ED respectively. Draw a plan of the remaining solid and the elevations in the vertical planes (a) parallel to AB, (b) parallel to AC. From the drawings, measure the angle that plane BXYC makes with the horizontal and the length of the line XC.'

Start the problem by making a rough sketch of the solid. In an example such as this, where a solid has been cut by a plane (truncated), it is better to draw the original solid first, after which the cutting plane can be drawn.

Notice that any edges which are hidden from view are shown by broken lines. The edges of the part of the solid which has been removed are also shown by broken lines.

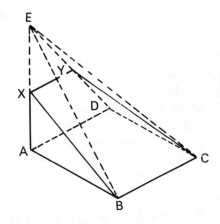

In drawing the plan and elevations, the small letters *a, b, c, d, x, y,* are taken to represent A, B, C, D, X, Y on the solid. The plan is drawn first and the elevations are drawn on the projection lines from the plan.

Scale: Three-quarter Size

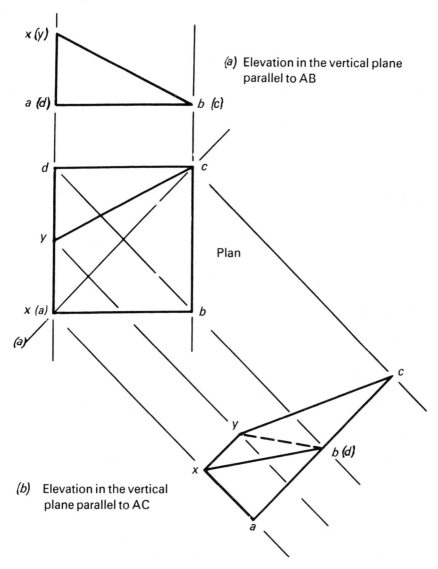

In drawing the plan, the following were considered:
(a) from the information, ABCD is a square of side 2". Since E is vertically above A, so is X, thus the line *ax* appears as a point on the plan.
(b) Y is the mid-point of ED, but as E was vertically above A, the edge ED is shortened to the line *ad* on the plan. Thus *y* is the mid-point of *ad*.

In drawing the elevation parallel to AB, the following were considered:
(a) A, B and X are all in the plane of the elevation, so the lines *ab, ax* and *xb* appear full size on the elevation.
(b) since XY, AD and BC are all parallel to the line of vision of the draughtsman, the points *y, d* and *c* will be obscured by *x, a* and *b*, respectively, on the elevation.

In drawing the elevation parallel to AC, the following were considered:
(a) projection lines perpendicular to AC were needed, and so were drawn.
(b) since AX is in the plane of the elevation, *ax* appears full size on the elevation.
(c) *yd* is shown as a broken line. This indicates that YD would not be visible from the line of vision of the draughtsman.

To measure the angle that plane BXYC makes with the horizontal:
The line *bx* on the elevation parallel to AB gives the line of slope of the plane. Thus *abx* is the required angle.
By measurement, the plane BXYC makes an angle of $26\frac{1}{2}°$ with the horizontal.

To measure the length of the line XC:
This line is contained in the plane of the elevation parallel to AC, thus the length of XC is equal to the length of *xc* on this diagram.
By measurement, (using dividers and ruler), the length of XC is 3".

The Terms, Front, End and Side Elevations
These terms are used when making drawings of everyday objects where we are so familiar with the object that we are immediately sure what the *front, end* or *side* is. (e.g. a house, a book, a chair, etc.) The elevations are taken in planes parallel to the front, end or side as desired.

Observation
Above all, the student must develop his powers of observation when dealing with this topic. The aim of this chapter and the test which follows is to improve the ability of the student at being able to predict what should (*and should not*) be in his finished drawing.

Class Test: Plan and Elevation
40 minutes 20 questions

1.

Front Elevation 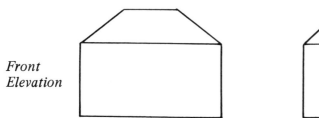 *End Elevation*

The above drawings are the front and end elevations of a house with a hipped roof. Which of the following is the plan of the house?

A.  B.

C. 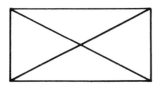 D.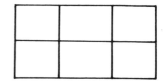

E.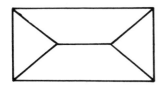

2. The drawing below includes the plan and elevation of a building with a spire.

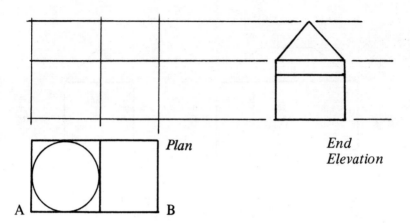

Which of the following is the front elevation of the building in the vertical plane parallel to AB?

A.

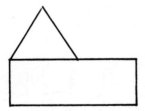

B.

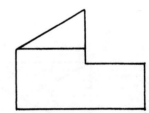

C.

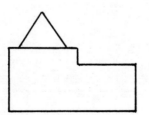

D.

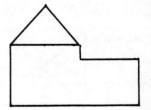

E.

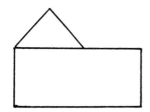

3.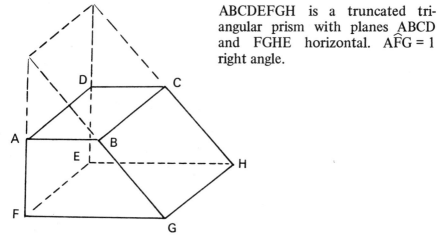

ABCDEFGH is a truncated triangular prism with planes ABCD and FGHE horizontal. $\widehat{AFG} = 1$ right angle.

If ABCD is a square, which of the following is the elevation of the solid in the vertical plane parallel to AC?

A.

B.

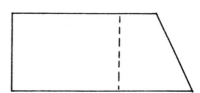

C.
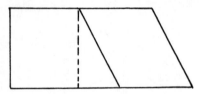

D.
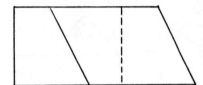

E.

4. A plan is drawn of a right square based pyramid. Which of the following statements is not true?
   A. The boundary (or perimeter) of the figure would be a square.
   B. The position of the vertex would be visible on the figure.
   C. The true length of a sloping edge of the pyramid could be measured on the figure.
   D. It would be impossible to measure the height of the vertex above the base on the figure.
   E. All the edges of the pyramid would be visible on the figure.

5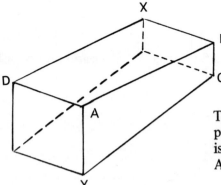

The diagram shows a view of a right prism whose uniform cross-section is a trapezium (e.g. ABCY, where AY ∥ BC).

If it is required to find the length of XY by direct measurement, on which of the following drawings could this be done?
   A. The plan of the solid.

B. The elevation of the solid in the vertical plane parallel to AB.
C. The elevation of the solid in the vertical plane parallel to AD.
D. The elevation of the solid in the vertical plane parallel to AX.
E. The view of the solid as given here.

6.

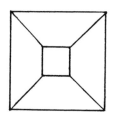

The diagram is a plan of a solid. Four of the solids, of which views are given below, would have this diagram as their plan. Which one would not?

A.

B.

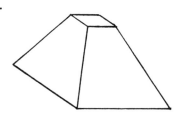

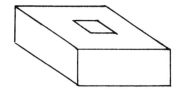

C.

D.

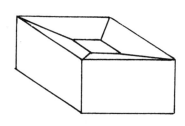

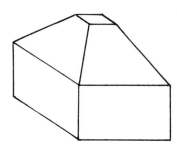

E.

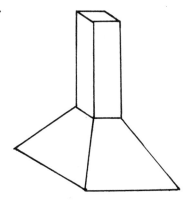

(*In each diagram the small square is placed centrally above the larger horizontal base square.*)

7.

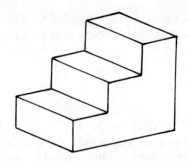

The accompanying diagram is a view of a short flight of stairs.

Which of the drawings given below is *not* an elevation of the solid in a vertical plane?

A.

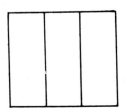

B.

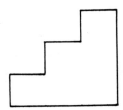

C.

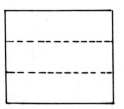

D.

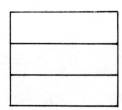

E.

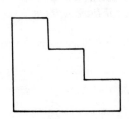

8. Each of the following drawings is accompanied by the name of a solid. Four of the drawings show elevations in a vertical plane of the solid mentioned; one does not. Which one?

A.

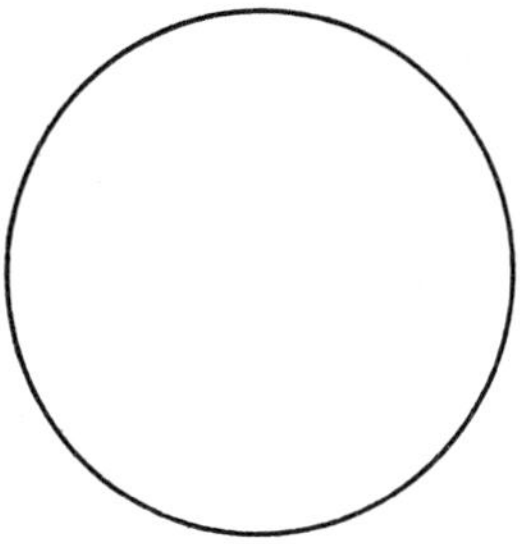

A sphere

B.

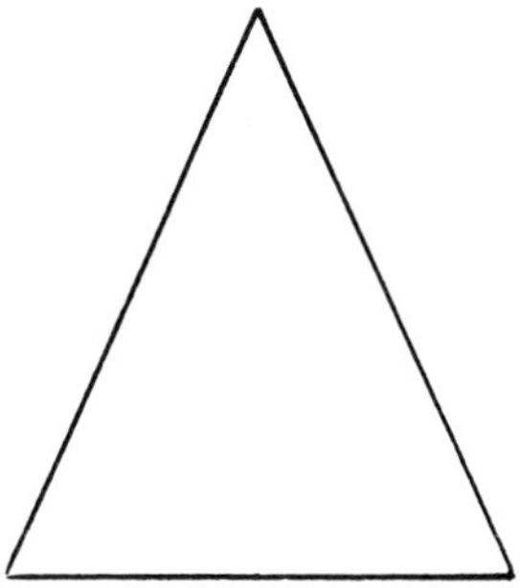

A right circular cone

C.

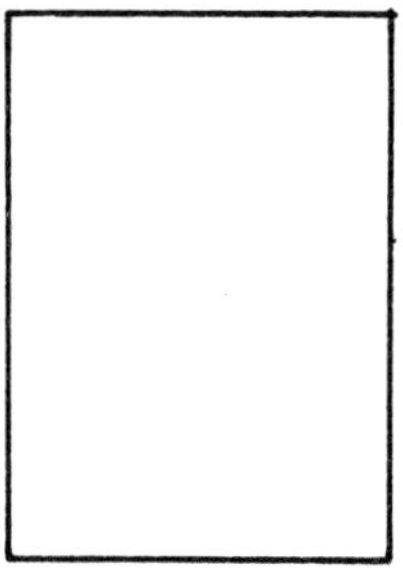

A cube

D.

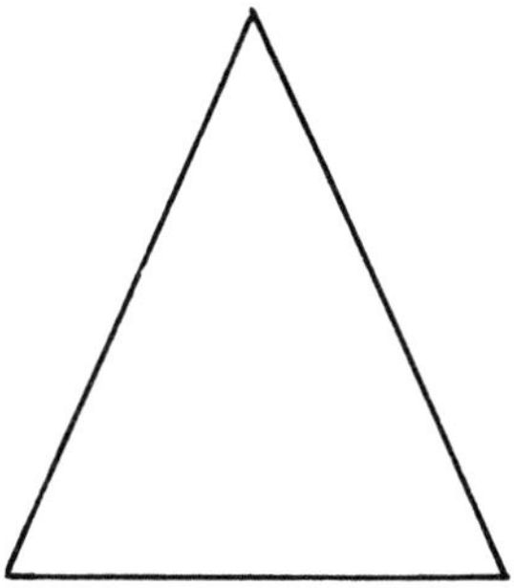

A right pyramid

E.

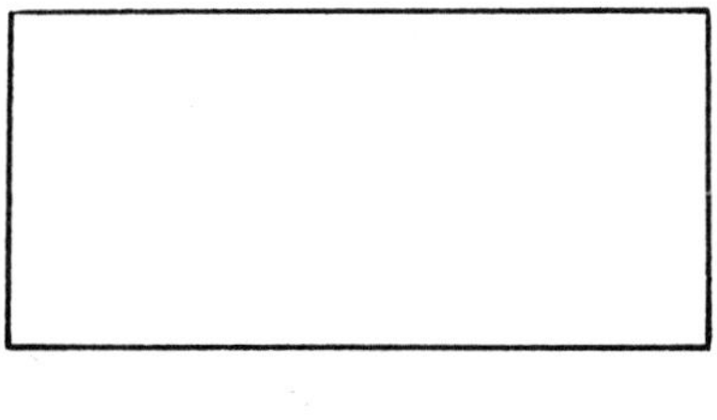

A cuboid

101

9. The drawings below are the plan and front elevation of a mosque with two domed minarets from which people are called to prayer.

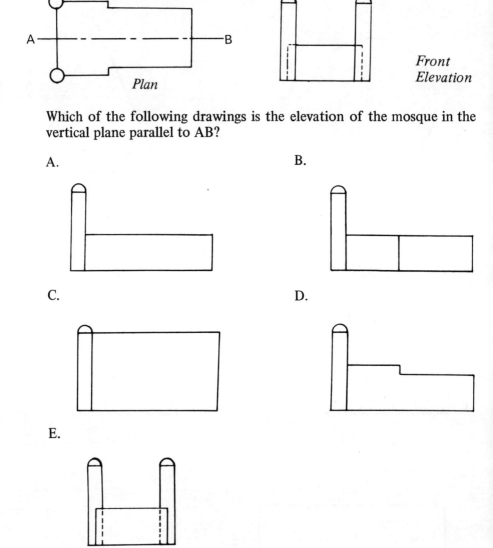

Which of the following drawings is the elevation of the mosque in the vertical plane parallel to AB?

10. The vertex of a tetrahedron (solid with four triangular faces) is vertically above the mid-point of one of the sides of its base. If its base is an equilateral triangle, which of the following drawings is the plan of the tetrahedron?

A.

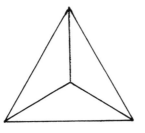

B.

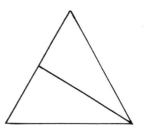

C.

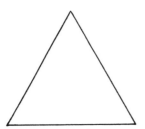

D.

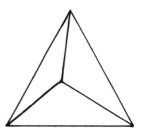

E.

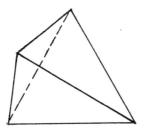

Use the following drawings and information to answer questions 11, 12, 13.

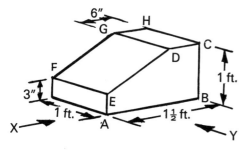

The diagram shows a view of a foot rest, such as is found in a shoe shop. It is in the form of a truncated cuboid and the dimensions are given on the diagram. On p. 104 drawn an accurate plan and the front (X→) and side (Y→) elevations of the solid.

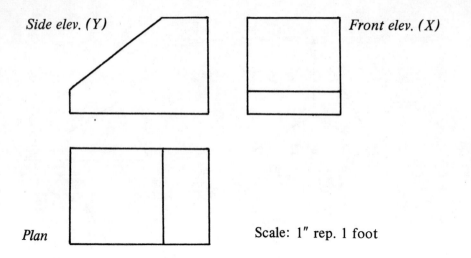

Scale: 1" rep. 1 foot

You will need dividers, a ruler and a protractor to answer the questions.

11. Find, by measuring on the correct *scale* drawing, the actual distance of corner A from corner C.

   A.   1·8 in.

   B.   1 ft. 9·6 in.

   C.   1 ft. 6 in.

   D.   1 ft. 7·8 in.

   E.   1·5 in.

12. What is the inclination of plane DEFG to the horizontal?

   A. 27°

   B. 30°

   C. 37°

   D. 43°

   E. 45°

13. What is the distance of corner A from corner H?

   A. 2 ft.

B. 1 ft. 9·6 in.

C. 1 ft. 6 in.

D. 3 ft. 6 in.

E. This distance cannot be determined from the given drawings.

14. A woodwork class was given a block of wood and the following two drawings of the plan and elevation of a solid.

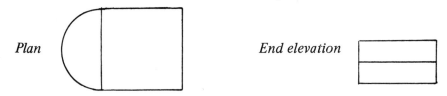

The class was told to make a solid which had the given plan and elevation. Here are sketches of the models that three boys made:

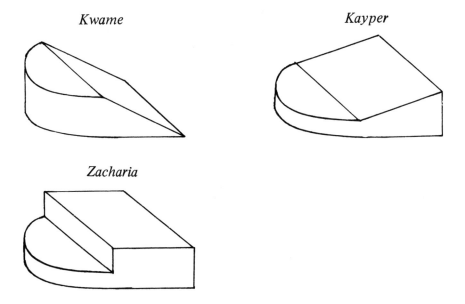

Which boy(s) had constructed a correct model?

    A. Kwame only

    B. Zacharia only

C. Kayper and Zacharia

D. Kwame and Kayper

E. Kwame and Zacharia

15. Drawn below are the end elevations of four common objects:

Which of the following is not represented above?

A. Fly-spray

B. Blackboard ruler

C. Book

D. Blackboard eraser

E. Pencil

16. Drawn below are the plans of four common objects:

 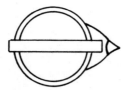 

Which of the following is not represented above?

A. Battery

B. Filter funnel

C. Sunglasses

D. Kettle

E. Torch

17.

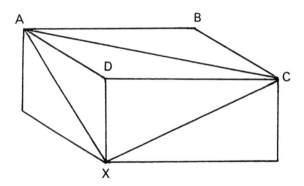

The above cuboid is truncated by a plane which passes through the points A, C and X. Which of the following is an elevation of the remaining solid in the vertical plane parallel to edge AB? (ACXD having been removed.)

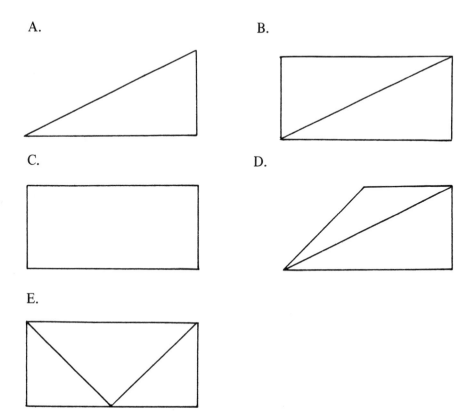

18.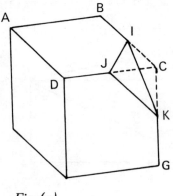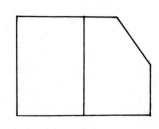

Fig. (a) *Fig. (b)*

Fig. (a) is a view of a truncated cube. The cube has been truncated by a plane passing through I, J and K, the mid-points of its edges BC, CD and CG respectively.
Fig. (b) is an elevation in the vertical plane parallel to . . .

A. AC

B. AI

C. AD

D. AJ

E. None of these

19. Drawn below are the plan, front elevation and end elevation of a solid:

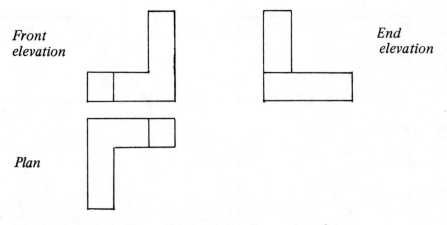

Which of the following solids has been drawn above?

A.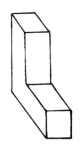

B.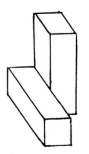

C.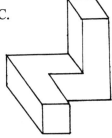

D.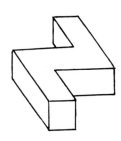

E.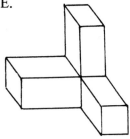

20. As in question 19, the drawings below are the plan, front elevation and end elevation of a solid:

Front elevation *End elevation*

Plan

Which of the following solids has the above plan and elevations?

A.

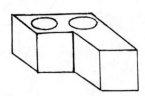

B.

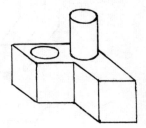

C.

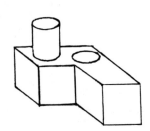

D.

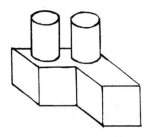

E. Solids A, B, C and D all have the given plan and elevations.

Chapter 10 Plane Geometry

The student should be familiar with the common angle and line properties of the following:
 Adjacent and vertically opposite angles
 Parallel lines and a transversal
 The different triangles
 The different quadrilaterals
 Polygons
 Circles
 Tangents to a circle
Knowledge of the concepts of similarity and congruence is essential.

Objective Test Methods
Objective questions in plane geometry can be answered much more quickly than written questions. There is no need to write out a proof, mentioning the theorems used. All the work can be done on the given diagram.

Angle Problems
While there is usually a quick and neat approach to solving an angle problem in geometry, weaker students may get better, if slower, results if they steadily work their way round the diagram, writing in the sizes of any angles they know. In this way the required angle will eventually be found. To aid the recognition of angle properties the student should memorize simple sketch diagrams which represent the important theorems. This has much more meaning and use than the theorem stated in words. Thus, each of the following sketches is a pictorial representation of a theorem, or of an axiom:

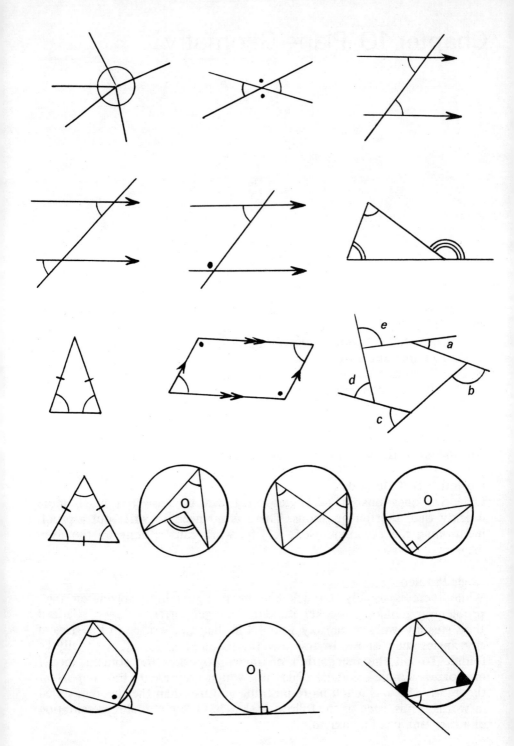

Comparative Lengths of Straight Lines

Watch out for sets of parallel lines (mid-point theorem, equal intercept theorem), similar triangles and the special quadrilaterals (some of which are shown below).

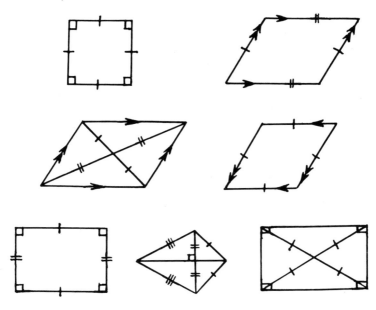

Two radii, drawn within the same circle give an isosceles triangle which is often useful.

Tangents from a point outside a circle to the circle are equal in length.

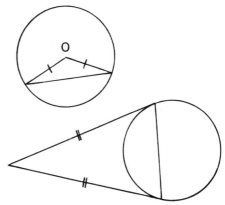

Common Error

Do not assume that an angle which looks like a right angle is a right angle. Similarly for angles which are apparently equal to 45° and lines which appear to be equal in length.

Class Test: Plane Geometry
40 minutes 20 questions

1. In the diagram, $\overline{AF}$, $\overline{BE}$, $\overline{CG}$ intersect at D and $\overline{AB} \parallel \overline{CG}$. If $D\hat{A}B = 113°$ and $C\hat{D}E = 22°$, what is the size of $A\hat{D}B$?

    A. 22°
    B. 45°
    C. 67°
    D. 89°
    E. 91°

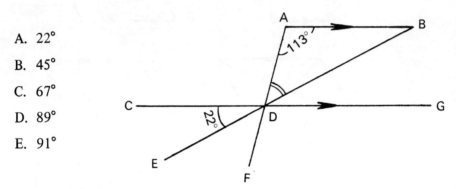

2. In the diagram, $D\hat{A}B$ and $E\hat{B}A$ are two exterior angles of $\triangle ABC$. If $D\hat{A}B = 82°$ and $A\hat{C}B = 51°$, what is the size of $A\hat{B}E$?

    A. 82°
    B. 98°
    C. 113°
    D. 133°
    E. 149°

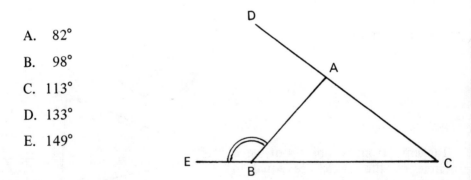

3. In the diagram, a square BCEF has been drawn within the rectangle ABCD. The diagonals of the rectangle intersect at X, similarly the square's diagonals intersect at Y. If |AB| = 5" and |BC| = 4", what is |XY|?

114

A. ¼"
B. ½"
C. 1"
D. 2"
E. 2½"

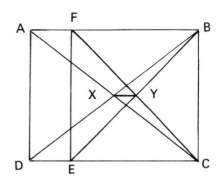

4. ABCD is a parallelogram. $\overline{XB}$ bisects $A\hat{B}C$ and $B\hat{X}C$ is a right angle. If $A\hat{D}C = 118°$, what is $B\hat{C}X$?

   A. 118°
   B. 62°
   C. 59°
   D. 45°
   E. 31°

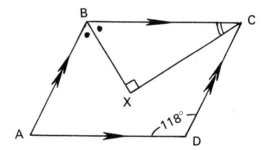

5. In trapezium PQRS, $\overline{QR} \parallel \overline{PS}$ and $|PQ| = |SR|$. If $P\hat{S}R = 55°$, what is the size of $P\hat{Q}R$?

   A. 55°
   B. 110°
   C. 125°
   D. 145°
   E. More information is needed.

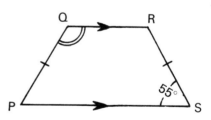

6. In the diagram, $\overline{TA}$ and $\overline{TB}$ are tangents to the circle, centre O. Beside the diagram, four triangles are named. Which of those triangles is (are) isosceles?

(1) △ TAO

(2) △ TOB

(3) △ ATB

(4) △ AOB

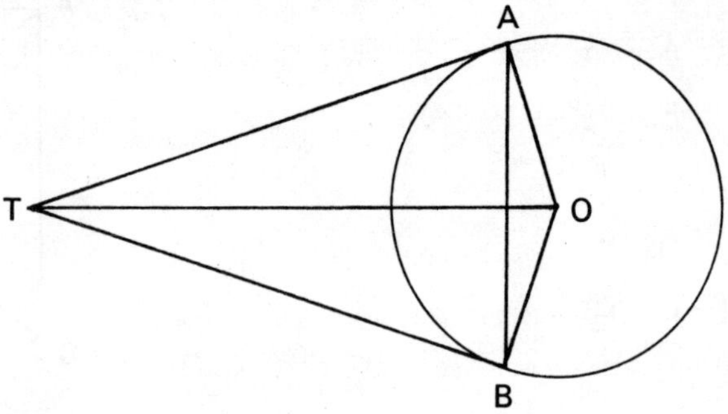

A. Both (3) and (4)

B. Both (1) and (2)

C. (4) only

D. All the triangles are isosceles.

E. None of the triangles is isosceles.

7. A large pair of scissors is left lying open on a table. Using the dimensions given in the diagram, find the distance, d, between the tips of the blades.

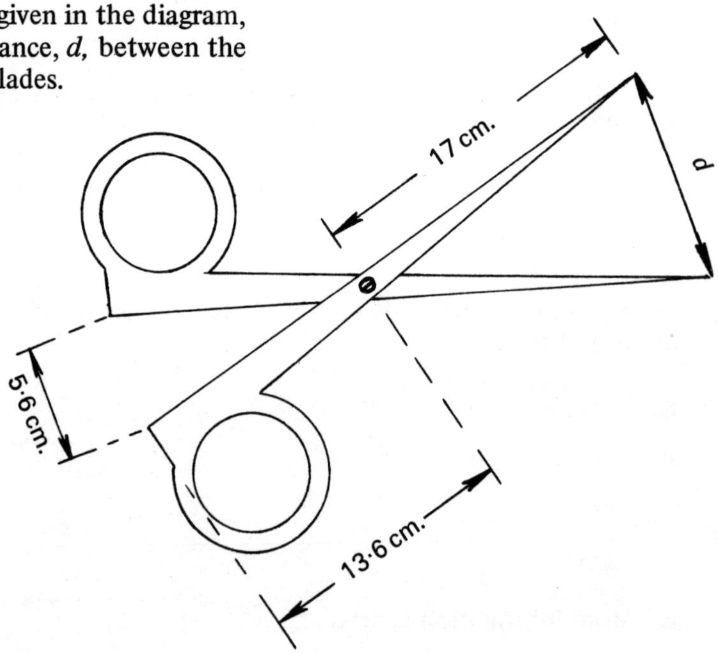

A. 5·6 cm.

B. 6·8 cm.

C. 7 cm.

D. 8 cm.

E. 9 cm.

8. A square is inscribed in a circle. Which statement is not true?
   A. The diagonals of the square intersect at the centre of the circle.
   B. A diagonal of the square will be a diameter of the circle.

C. Each side of the square subtends one right angle at the centre of the circle.
D. Each side of the square subtends one right angle at the circumference of the circle.
E. The angle subtended by a side of the square in the minor segment is 135°.

9. ABCDEF is a regular hexagon. A point G is taken outside the hexagon so that |GA| = |GE| = |AE|. What is the size of reflex angle BAG?

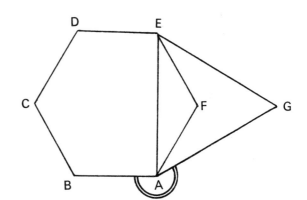

A. 120°
B. 150°
C. 210°
D. 240°
E. 300°

10. Points P, Q, R, S, T are taken on the circumference of a circle so that $\overline{QR} \parallel \overline{PS}$, $Q\hat{S}T = 87°$ and $R\hat{Q}S = 23°$. What is the size of $P\hat{Q}T$?

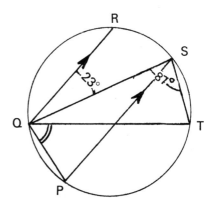

A. 45°
B. 46°
C. 60°
D. 64°
E. 67°

117

11. O is the centre of circle RST, $R\hat{S}T = 73°$ and $S\hat{O}T = 62°$. What is $R\hat{T}O$?

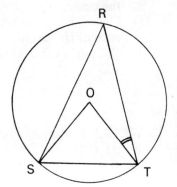

A. 11°
B. 14°
C. 15½°
D. 17°
E. 31°

12. In the diagram, O is the centre of circle ABCD in which the arcs AB and BC are equal in length. If $A\hat{D}B = 26°$, what is the size of $D\hat{O}C$?

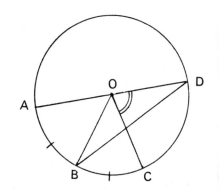

A. 52°
B. 76°
C. 90°
D. 104°
E. 128°

13. In the diagram, $\overline{TS}$ is a tangent to a circle at S and $\overline{TQ} \parallel \overline{SP}$. If $S\hat{Q}R = x°$ and $P\hat{Q}S = y°$, what is $R\hat{T}S$?

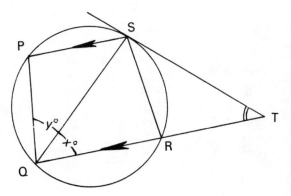

A. $x°$
B. $y°$
C. $(x + y)°$
D. $180° - (x + y)°$
E. $(90 - x)°$

14. PQRST is a pentagon in which $\overline{PQ} \parallel \overline{TS}$, $\hat{Q} = 129°$, $\hat{R} = 4x°$ and $\hat{S} = 3x°$. What is the value of x?

   A. 17
   B. 30
   C. 33
   D. 51
   E. More information is needed.

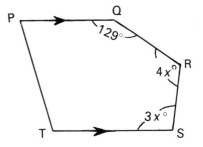

15. In the diagram, $\overline{EC}$ is a diameter of circle ABCDE. If $A\hat{B}C = 158°$, what is the size of $A\hat{D}E$?

   A. More information is needed.
   B. 112°
   C. 90°
   D. 68°
   E. 22°

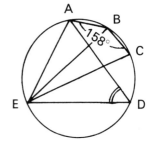

16. In the diagram, A, B, C, D, E are points on the circumference of the circle. $\overline{AD}$ and $\overline{CE}$ intersect at X. If $E\hat{B}D = 36°$ and $A\hat{D}C = 45°$, what is $A\hat{X}C$?

   A. 81°
   B. 99°
   C. 72°
   D. 108°
   E. 90°

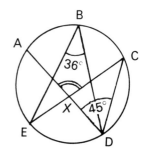

17. Which of the following figures does *not* contain a pair of similar triangles?

I II

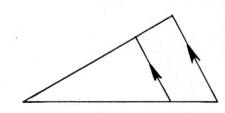

III IV

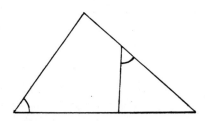 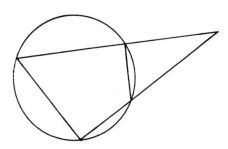

A. I only

B. II only

C. III only

D. IV only

E. Each figure contains a pair of similar triangles.

18. $\overline{TA}$ is a tangent to the circle ABC. $\overline{TA} \parallel \overline{CB}$ and D is a point outside the circle so that $\overline{BD} \parallel \overline{AC}$. In what order must steps I-V be placed in proving $\overline{CB}$ bisects $A\hat{B}D$?

   I. $T\hat{A}C = A\hat{B}C$ (alt. segment th.)

   II. $C\hat{B}D = A\hat{C}B = T\hat{A}C = A\hat{B}C$

   III. $A\hat{C}B = T\hat{A}C$ (alt. angles, $\overline{TA} \parallel \overline{TB}$)

   IV. $\overline{CB}$ bisects $A\hat{B}D$ ($C\hat{B}D = A\hat{B}C$, shown above)

   V. $C\hat{B}D = A\hat{C}B$ (alt. angles, $\overline{AC} \parallel \overline{BD}$)

A. I, II, III, IV, V

B. I, II, III, V, IV

C. II, V, III, I, IV

D. V, III, I, II, IV

E. V, I, II, III, IV

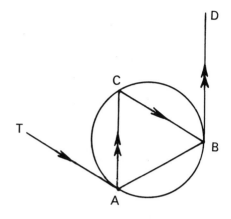

19. The diagram shows two unequal circles intersecting at X and Y. AXB and DYC are straight lines. Which of the following statements is true?

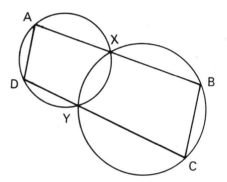

A. ABCD is a cyclic quadrilateral.

B. ABCD is a trapezium.

C. ABCD is a parallelogram.

D. ABCD is a square.

E. ABCD is a rectangle.

20. Four well known triangle theorems are stated below.

1. The sum of the interior angles of a triangle is two right angles.

2. In a right angled triangle, the sum of the squares on the sides containing the right angle is equal to the square on the hypotenuse.

3. An exterior angle of a triangle is equal to the sum of the two opposite interior angles.

4. The line joining the mid-points of two sides of a triangle is parallel to, and equal to half the length of, the third side.

The following sketches could serve as reminders for triangle theorems. The theorem connected with one of the sketches has not been quoted. Which one?

A. B. C.

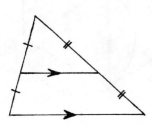 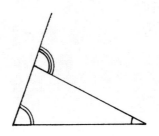 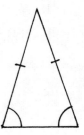

D. E.

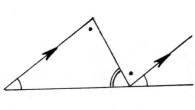 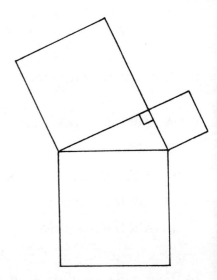

Chapter 11 Plane Trigonometry

The student should be familiar with the trigonometrical ratios, sine (sin), cosine (cos) and tangent (tan), of the angles in a right angled triangle. Knowledge of the inverse ratios, cosecant (cosec), secant (sec) and cotangent (cot), is also useful. Using these ratios, he should be able to solve right-angled triangles, and, using the Sine rule and the Cosine rule, he should be able to solve triangles which do not contain a right angle.

Solving Triangles
(By *solving* a triangle, we mean finding the lengths of its sides and the sizes of its angles.)
    It is important to be able to solve triangles as applications are found in many fields outside mathematics: map-making, map-reading, surveying, mechanics, navigation, etc.
    Right-angled triangles can be solved completely by applying the following:
(a) the 'angle sum of triangle' theorem,
(b) the appropriate trigonometrical ratio.
    Triangles which are not right-angled can be solved by applying one or more of the following:
(a) the 'angle sum of triangle' theorem,
(b) the Sine rule,
(c) the Cosine rule.

The Trigonometrical Ratios of the Angles 45°, 30°, 60°
The angles 45°, 30°, 60°, have simple trigonometrical ratios which should be used where possible.

45°
Consider an isosceles right-angled triangle whose equal sides are each of length 1 unit. The length of the hypotenuse is $\sqrt{2}$ units (Pythagoras' theorem).
From the diagram,

$\sin 45° = \dfrac{1}{\sqrt{2}}$

$\cos 45° = \dfrac{1}{\sqrt{2}}$

$\tan 45° = 1$

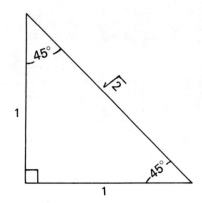

30°, 60°
Consider an equilateral triangle, ABC, of length of side 2 units. $\overline{AD}$ is the perpendicular from A to $\overline{BC}$. Thus, $B\hat{A}D = 30°$ and $|BD| = 1$ unit (symmetry about axis AD) and $|AD| = \sqrt{3}$ units (Pythagoras' theorem).
From △ABD,

$\sin 30° = \tfrac{1}{2}$

$\cos 30° = \dfrac{\sqrt{3}}{2}$

$\tan 30° = \dfrac{1}{\sqrt{3}}$

$\sin 60° = \dfrac{\sqrt{3}}{2}$

$\cos 60° = \tfrac{1}{2}$

$\tan 60° = \sqrt{3}$

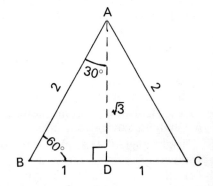

Instead of trying to commit these ratios to memory, memorize the diagrams and derive the appropriate ratio when it is needed.

Trigonometrical Ratios of Obtuse Angles
The following should be remembered:
(In each case, $\theta°$, is an obtuse angle.)
(a) $\sin \theta° = \sin(180 - \theta)°$
(b) $\cos \theta° = -\cos(180 - \theta)°$
(c) $\tan \theta° = -\tan(180 - \theta)°$

Class Test: Plane Trigonometry
40 minutes 20 questions

Use the following figures and information to answer questions 1, 2, 3, 4, 5, 6.
The figures show four triangles ABC, each containing a right angle and two sides whose lengths are x and y units.

1.

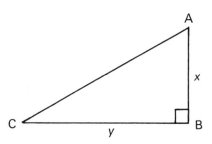

2.

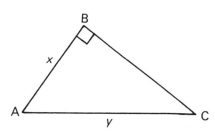

3.

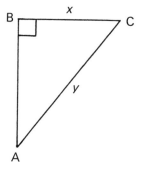

4.

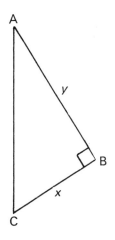

1. In which triangle does $\tan \widehat{C} = \dfrac{x}{y}$?

   A. 1.

   B. 2.

   C. 3.

   D. 4.

   E. Neither 1., 2., 3. nor 4.

2. In which triangle does $\sec \widehat{C} = \dfrac{y}{x}$?

   A. 1.

   B. 2.

   C. 3.

   D. 4.

   E. Neither 1., 2., 3. nor 4.

3. In which pair of triangles does $\sin \widehat{A} = \tan \widehat{A} = \dfrac{x}{y}$?

   A. 1. and 2.

   B. 2. and 3.

   C. 3. and 4.

   D. 1. and 3.

   E. 2. and 4.

4. In which pair of triangles does $\cos \widehat{A} = \cot \widehat{C} = \dfrac{x}{y}$?

   A. 1. and 2.

   B. 2. and 3.

   C. 3. and 4.

   D. 1. and 3.

   E. 2. and 4.

5. Umaru, Gideon and Abdul were discussing the triangles:
   Umaru: The four triangles are similar.
   Gideon: Triangle 1. is congruent to triangle 4.; also, triangles 2. and 3. are congruent.
   Abdul: $\widehat{A}$ in triangle 1. is not equal to $\widehat{A}$ in triangle 2.
   Which boy(s) is (are) correct?

   A. Umaru only

   B. Gideon only

   C. Abdul only

D. Gideon and Abdul are correct.

E. Umaru and Gideon are correct.

6. In which three triangles does $\cos \widehat{C} = \sin \widehat{C} = \cot \widehat{A} = \dfrac{x}{y}$?

   A. 1., 2. and 3. respectively

   B. 3., 2. and 1. respectively

   C. 2., 3. and 4. respectively

   D. 3., 2. and 4. respectively

   E. 2., 1. and 4. respectively

7. Which of the following statements is incorrect if angles a, b, c, d, e all lie between $0°$ and $180°$?

   A. $\tan a = 2{,}500$

   B. $\cos b = 0\cdot7431$

   C. $\cos c = -0\cdot8116$

   D. $\sin d = 0\cdot6616$

   E. $\sin e = -0\cdot9163$

8. In the diagram, ABC is a straight line and D is such that $\cos A\widehat{B}D = \frac{1}{2}$. What is $\sin D\widehat{B}C$?

   A. $\sqrt{3}$

   B. $\dfrac{\sqrt{3}}{2}$

   C. $\frac{1}{2}$

   D. $-\frac{1}{2}$

   E. $-\dfrac{\sqrt{3}}{2}$

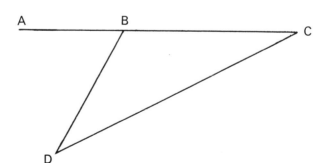

9. In the diagram, O is the centre of the circle whose radius is r. The chord AB subtends an angle of 2θ at O. What is $|AB|$ in terms of r and θ?

A. $r \sin(180° - 2\theta)$

B. $r \sin 2\theta$

C. $2r \sin 2\theta$

D. $r \sin \theta$

E. $2r \sin \theta$

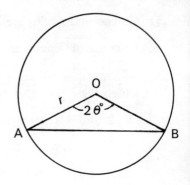

10. The transversal joining the equal uprights of a letter 'N' is 1 inch long. If the angle between the transversal and an upright is 45°, what is the length of an upright?

    A. $\sqrt{2}$ in.

    B. $\dfrac{2}{\sqrt{2}}$ in.

    C. 1 in.

    D. $\dfrac{1}{\sqrt{2}}$ in.

    E. $\tfrac{1}{2}$ in.

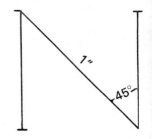

11. In $\triangle ABC$, $\sin \widehat{A} = \tfrac{3}{4}$, $|BC| = 18''$ and $|AC| = 12''$. What is $\sin \widehat{B}$?

    A. $\tfrac{1}{2}$

    B. $\tfrac{2}{3}$

    C. $\tfrac{8}{9}$

    D. $\tfrac{3}{2}$

    E. 2

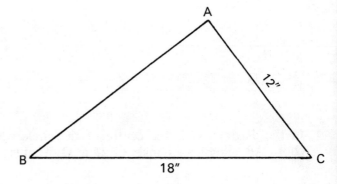

128

12. In △LMN, cos M̂ = ⅓, |LM| = 10″ and |MN| = 15″. What is |LN|?

    A. 20·6″
    B. 15″
    C. 11·2″
    D. 8·7″
    E. 8⅓″

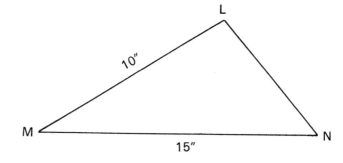

13. In △PQR, |PQ| = 6″, |QR| = 11″ and |RP| = 7″. What is cos P̂?

    A. ³⁄₇
    B. ⁶⁄₁₁
    C. ⁷⁄₁₁
    D. −³⁄₇
    E. −⁶⁄₇

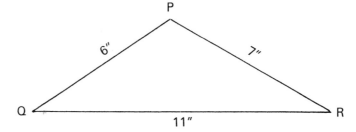

14. In △XYZ, X̂ = 44°, Ẑ = 36° and |YZ| = 5″. Which of the following will give |XZ| in inches?

    A. $\dfrac{5 \sin 100°}{\sin 44°}$

    B. $\dfrac{5 \sin 44°}{\sin 80°}$

    C. $\dfrac{5 \sin 36°}{\sin 44°}$

    D. $\dfrac{5 \sin 44°}{\sin 100°}$

    E. $\dfrac{5 \sin 80°}{\sin 36°}$

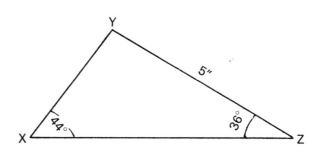

15. In △PQR, $\hat{Q} = \theta$ and $\hat{R} = \phi$. $\overline{PS}$ is the perpendicular from P to $\overline{QR}$ and |QS| and |SR| are a and b units respectively. Which of the following equations is true of the diagram?

   A. $\dfrac{a}{\sin \theta} = \dfrac{b}{\sin \phi}$

   B. $a \tan \theta = b \tan \phi$

   C. $a \cos \theta = b \cos \phi$

   D. $a \tan \phi = b \tan \theta$

   E. Neither A., B., C. nor D. are true.

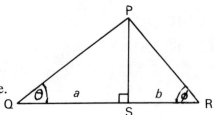

16. The angle of elevation of the top of a tree from the top of the head of a man, 6 feet tall, is 35°. If the man is standing on level ground at a distance of 50 feet from the tree, what is the height of the tree?

   A. (50 tan 55° + 6) feet

   B. (50 tan 55° − 6) feet

   C. (50 tan 35° + 6) feet

   D. (50 tan 35° − 6) feet

   E. (50 sin 35° + 6) feet

17. A lizard at the top of a 10 feet high vertical wall sees a beetle on the horizontal ground. If the angle of depression of the beetle from the lizard is 34°, how far is the beetle from the lizard?

   A. 10 sin 34° feet

   B. $\dfrac{10}{\sin 34°}$ feet

   C. 10 tan 34° feet

   D. 10 cos 56° feet

   E. $\dfrac{10}{\cos 34°}$ feet

18. In which of the triangles I, II, III, IV can |AB| be calculated using the equation, $|AB| = \dfrac{7 \sin 60°}{\sin 40°}$ inches?

I
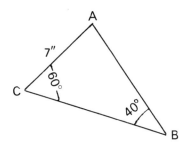

II
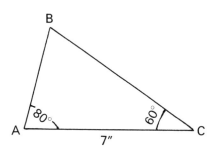

III
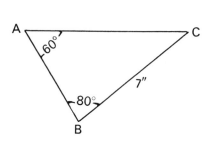

IV
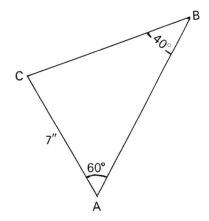

A. I and II
B. II and III
C. III and IV
D. I and III
E. I and IV

19. In which of the triangles I, II, III can $|PQ|$, in inches, be calculated by using the equation, $|PQ|^2 = 9^2 + 8^2 - 2 \times 8 \times 9 \times \cos 40°$?

I

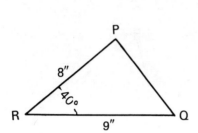

II

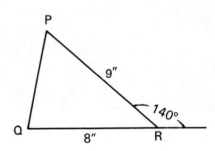

A. I only
B. III only
C. Both I and III
D. Both I and II
E. I, II and III

III

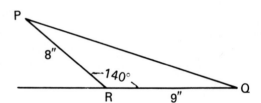

20. In the diagram, $B\hat{A}C$ is a right angle, $A\hat{B}C = 45°$ and $C\hat{D}A = 60°$. If $|AB| = 1''$, what is $|AD|$?

A. $\frac{1}{2}''$
B. $\frac{1}{\sqrt{3}}''$
C. $1''$
D. $\sqrt{3}''$
E. $2''$

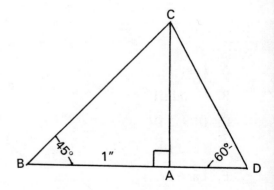

Chapter 12 Velocity Triangles

From given information, the student should be able to draw a *correctly arrowed* velocity triangle. To this end, he should be familiar with the meaning of the terms, *airspeed, windspeed, groundspeed, course, drift* and *track* (and the *nautical equivalents* of these). Either by trigonometrical methods, or from an accurate scale drawing, the student should be able to find from a velocity triangle the required speed and/or direction.

Vector Quantities
Velocity is a *vector quantity*. This means that to define the velocity of an object completely, its *speed and* the *direction* in which it is travelling must *both* be given. Thus, speed and velocity are not the same. The diagram shows what happens if 5 men leave a point O with equal speeds of 3 m.p.h., but in different directions.

The speeds are equal, but the velocities are different.

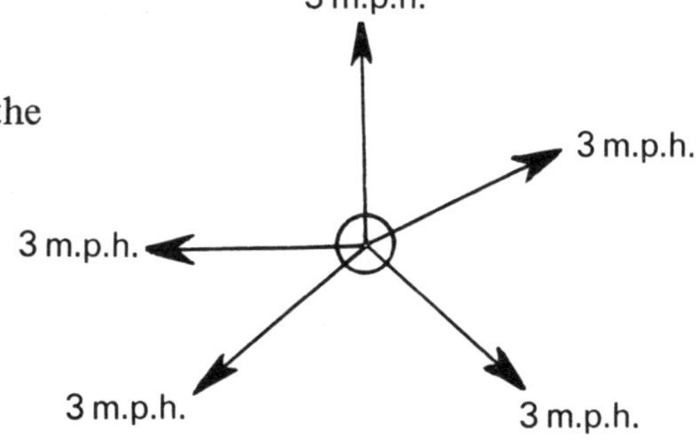

133

Any quantity which requires magnitude *and* direction before it is completely defined is known as a vector quantity. Other examples are *displacement, acceleration* and *force*.

Vector Addition
In the physics laboratory it can be demonstrated that two forces acting at a point can be replaced by a single resultant force. Further, the size and direction of this resultant force can be found by drawing the diagonal of a parallelogram whose adjacent sides represent the size and direction of the two original forces.

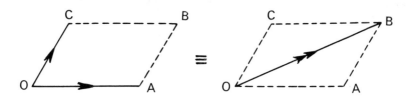

This can be represented in the following way: **OA + OC = OB**
This is a *vector equation,* where,
   OA represents the force acting along OA,
   OC represents the force acting along OC,
   OB represents the resultant of **OA** and **OC**, acting along OB.
   The above vector equation is an expression of a natural law known as the *Law of Vector Addition*. It has been shown to hold true for any vector quantity, and it states that *the sum of two vectors* **OA** *and* **AB** *is the vector* **OB**. (Note that this is a *vector sum, not an arithmetic sum*.)

In the above example, the same result is obtained from △OAB, thereby demonstrating the Law of Vector Addition as stated above. Note that as |AB| = |OC| and AB ∥ OC, then **AB = OC**.

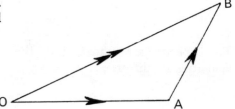

OA + AB = OB

If **OA** and **AB** are given, then the magnitude and direction of **OB** can be

found by calculating (or measuring in an accurately drawn triangle) |OB| and $A\hat{O}B$ in $\triangle OAB$.
Therefore, when dealing with velocity a correctly drawn and arrowed triangle will lead to the correct result.

Motion of an Aircraft

An aircraft is usually subjected to two velocities and the resultant of these gives its actual speed and direction over the ground (*ground velocity*). These velocities are:

(1) Air Velocity
This is a combination of the speed that the aircraft is capable of in still air (*airspeed*) and the direction set by the pilot of the aircraft (*course*).

(2) Wind Velocity
This is a combination of the speed of the wind and the direction *from* which it is blowing.
Thus, a typical velocity triangle for an aircraft's motion would be:

OA + AB = OB

Motion of a Ship

Similarly, the ground velocity (or velocity over the bed of the ocean) of a ship is the resultant of two velocity components:

(1) Velocity Through the Water
This is a combination of the speed that the ship is capable of in still water and the course set by the navigator.

(2) Current Velocity
A combination of the speed of the current and the direction in which it flows.

Thus, a typical velocity triangle for a ship's motion would be:

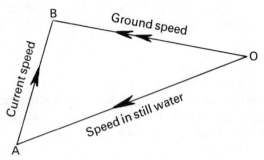

OA + AB = OB

Solving Velocity Triangles
Since a triangle contains three sides and three angles, navigation problems fall into three main groups:

(1) To Find the Ground Velocity
'An aircraft flies on a course of 052° with an airspeed of 200 m.p.h. If a N.W. wind is blowing at 50 m.p.h., draw a velocity triangle from which the groundspeed and the track could be found.'

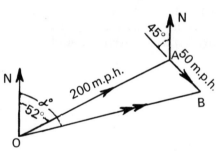

|OB| represents the groundspeed and the aircraft makes a track of $\alpha°$.

(2) To Find the Wind Velocity
'An aircraft makes good a track of 180° at a groundspeed of 150 m.p.h. If its airspeed is 180 m.p.h. and its course is 210°, draw a velocity triangle from which the windspeed and the direction from which it is blowing can be found.'

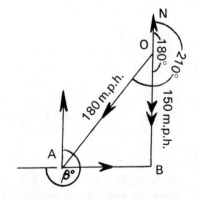

|AB| represents the windspeed and the wind is blowing from a direction $\beta°$.

(3) To Find the Speed in Still Water and the Course

'A ship is travelling at 20 knots in a direction N. 40° W. If the current flows at 4 knots in a direction N. 20° E., draw a velocity triangle from which the speed in still water and the course of the ship can be found.'

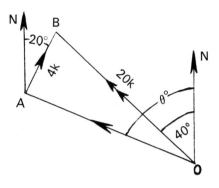

|OA| represents the speed in still water and the course is N. $\theta°$ W.

Study the above examples carefully and make sure that you understand how the triangles were obtained.

Class Test: Velocity Triangles
40 minutes 20 questions

Note: The arrowing of velocity triangles
The following method of arrowing is used on the diagrams to the questions:

Motion of Aircraft or Rain in a wind	Motion of ships or swimmers in moving water	Shown on the Velocity Triangle by
Airspeed Windspeed Groundspeed	Speed in still water Speed of the current Groundspeed	One arrowhead → One arrowhead → Two arrowheads ↠

The student may be familar with other methods of arrowing. This method emphasizes the point that the *velocity* of an aircraft in still air *taken with* the *velocity* of the wind *results in* the *true velocity* of the aircraft over the ground. (Similarly for ships in moving water.) An example of the arrowing is shown below:

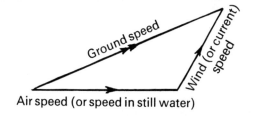

Use the given diagram and information to answer questions 1, 2, 3, 4.

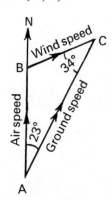

ABC is the velocity triangle for an aircraft's journey in a wind. |AB|, |BC|, |AC| represent the airspeed, windspeed and groundspeed respectively. $\overline{AB}$ lies on a line pointing north, $B\hat{A}C = 23°$ and $B\hat{C}A = 34°$.

1. What is the course of the aircraft?

    A. Due north

    B. N. 23° E.

    C. N. 57° E.

    D. S. 34° W.

    E. More information is needed.

2. What is the track of the aircraft?

    A. Due north

    B. N. 23° E.

    C. N. 57° E.

    D. S. 34° W.

    E. S. 57° W.

3. From what direction is the wind blowing?

    A. 034°

    B. 123°

    C. N. 57° E.

    D. S. 34° W.

    E. S. 57° W.

4. What is the drift of the aircraft?

    A.   34° to the left

    B. 123° to the left

    C.   23° to the right

    D.   23° to the left

    E. 123° to the right

5. Thinking carefully about the definitions of the words used in this topic, decide which of the following statements is incomplete.
    A. The speed of an aircraft in still air is 250 m.p.h.

B. The course of an aircraft was due south.
C. The drift of an aircraft was 45° to the right.
D. The wind velocity was 70 m.p.h.
E. The groundspeed of an aircraft was 280 m.p.h.

6. The course of an aircraft is given as 282°. Which of the following diagrams represents this information?

A.

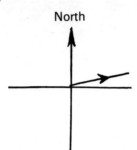

B.

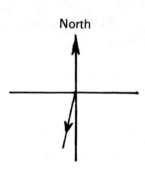

C.

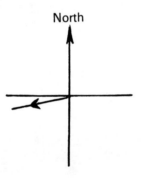

D.

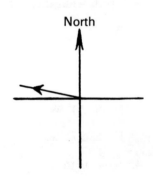

E.

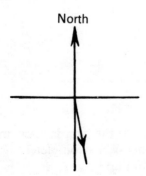

7. Rain falls vertically to the ground at 30 m.p.h. in still air. If a wind suddenly started to blow at 40 m.p.h. horizontally, what would be the speed of the rain?

   A. 30 m.p.h.
   B. 40 m.p.h.
   C. 50 m.p.h.
   D. 70 m.p.h.
   E. More information is needed.

8. . . . and what angle would the rainfall make with the ground?

   A. 30°
   B. arc tan $\frac{3}{4}$ (i.e. the angle whose tangent is $\frac{3}{4}$)
   C. 45°
   D. arc tan $\frac{4}{3}$
   E. 90°

9. A river flows downstream at 12 m.p.h. and a man can row in still water at 5 m.p.h. What would be the speed of the man if he rowed downstream?

   A. 5 m.p.h.
   B. 7 m.p.h.
   C. 12 m.p.h.
   D. 13 m.p.h.
   E. 17 m.p.h.

10. 'An aircraft can fly at 200 m.p.h. in still air. The pilot sets a course due north. Draw a sketch diagram to show the groundspeed, G, if there is a north-east wind blowing at 50 m.p.h.' Which of the following is the correct sketch diagram?

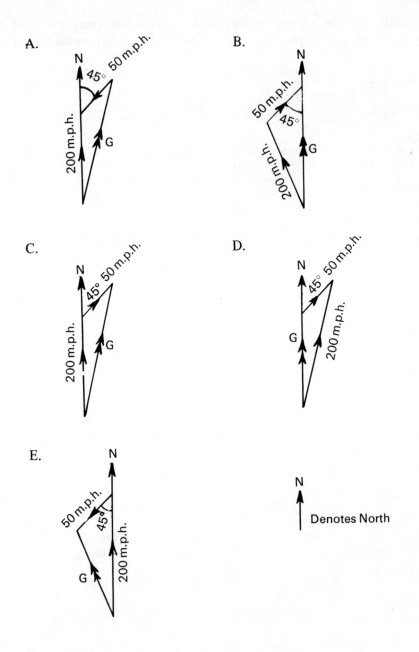

11. The diagram shows a velocity triangle, ABC. N_1 and N_2 point northwards. If angle $N_1 AB = 71°$ and reflex angle $N_2 CB = 312°$, what is $A\hat{B}C$?

A. 132°
B. 119°
C. 113°
D. 109°
E. 90°

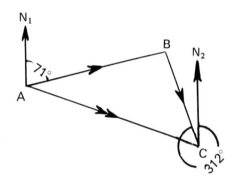

12. A man who can swim at 2 m.p.h. in still water is standing on one bank of a river which is flowing at 1 m.p.h. He wishes to swim in such a direction that he finishes on the other bank directly opposite his present position. What angle, ϕ, should his direction make with the bank he is standing on?

   A. It is impossible for the man to swim directly across the river.
   B. 30°
   C. 45°
   D. 60°
   E. 90°

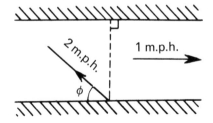

13. ... and what would be his resultant speed across the river?

   A. 3 m.p.h.
   B. $\sqrt{5}$ m.p.h.
   C. 2 m.p.h.
   D. $1\frac{1}{2}$ m.p.h.
   E. $\sqrt{3}$ m.p.h.

14. Which of the following triangles is a velocity triangle? (Remember that the groundspeed has two arrowheads.)

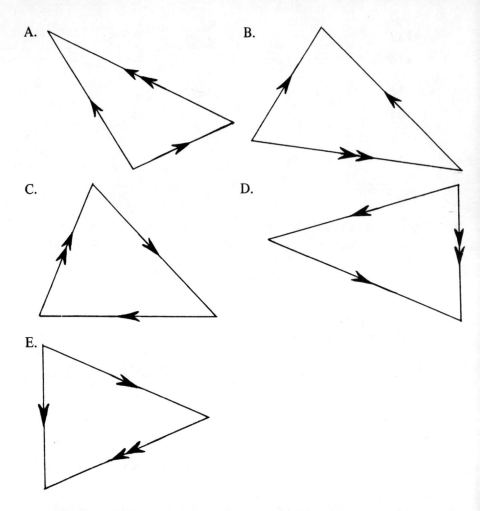

15. Which of the following statements is (are) correct?
    I. A south-west wind blows towards the north-east.
    II. If a man cannot swim faster than the speed of the current of a river, then he will not be able to swim across the river.
    III. In a velocity triangle, the longest side always represents the groundspeed.

    A. I only

    B. II only

    C. III only

    D. Both I and II

    E. Both I and III

16. The diagram shows a velocity triangle. Two of its sides represent speeds of v and $2v$ and the angle between those sides is 120°. What is the groundspeed, G, expressed in terms of v?

    A. $v\sqrt{7}$

    B. $v\sqrt{5}$

    C. $v\sqrt{3}$

    D. $\dfrac{3v}{2}$

    E. $3v$

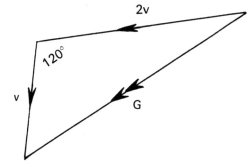

17. A pilot flies the best possible course between two airports which are 1000 miles apart. His airspeed is 200 m.p.h., his groundspeed is 240 m.p.h. and the wind blows from a fixed direction at 50 m.p.h. What is the time for the journey?

    A. 20 hours

    B. 5 hours

    C. 4 hours 10 minutes

    D. 4 hours

    E. 2 hours 16 minutes

18. B is due north of A. A pilot flies from A to B on a course N. 20° W. and then makes the return flight from B to A, maintaining the same airspeed during both flights. If the wind blows at a constant speed from the south-west during both flights, which of the following statements is (are) correct?
    I. The groundspeed will be the same during each flight.
    II. The pilot's course for the return flight will be S. 20° W.
    III. The return flight will take longer than the outward flight.

    A. I only

    B. II only

    C. III only

    D. Both I and II

    E. Both II and III

19. A ship travels with an engine speed of 20 knots on a course of 210°. The navigator calculates that the ship is actually making 17 knots in a direction 168°. If the ship is travelling in an ocean current of constant speed and direction, which of the following sketches represents the above information?

A.

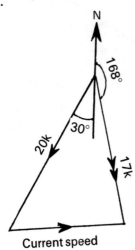

B.

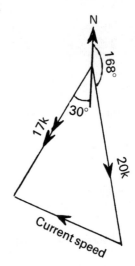

C.

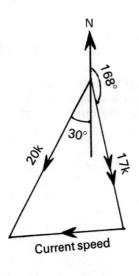

D.

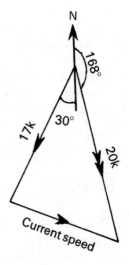

E.

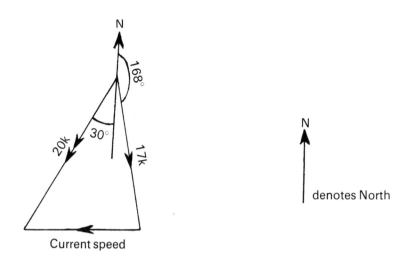

Current speed

denotes North

20. An aircraft, flying in a wind which is blowing at 45 m.p.h. from S. 52° W., makes good a track N. 12° W. If its groundspeed is 340 m.p.h., which of the following velocity triangles represents this information?

A.

B.

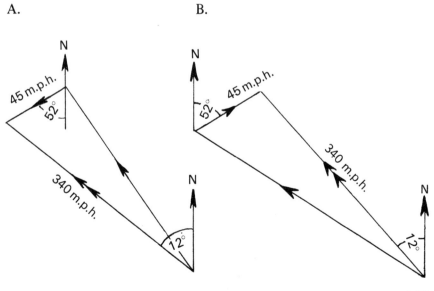

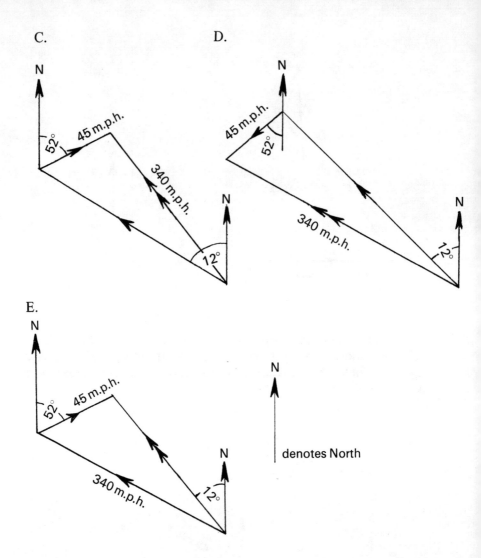

Part II: Certificate-level Tests

Test 1
75 minutes 50 questions

1. In an objective test a student must attempt 50 questions in 1¼ hours. What is the average time that he should spend on each question?

    A. ⅔ of a minute
    B. 1 min.
    C. 1¼ min.
    D. 1½ min.
    E. 3 min.

2. In the diagram, $\overline{TA}$ and $\overline{TB}$ are tangents to the circle. |TA| = 12" and |TD| = 5". What is |DB|?

    A. 60"
    B. 17"
    C. 8½"
    D. 7½"
    E. 7"

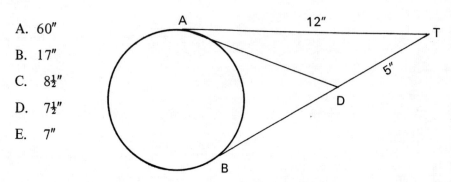

150

3. Simplify $3(5a^2 + 2c) - 2a(1 - 3a) - 6c$.

   A. $21a^2 - 2a - 6c$
   B. $13a^2 - 2a - 12c$
   C. $9a^2 - 2a$
   D. $13a^2 - 2a - 6c$
   E. $21a^2 - 2a$

4. What is the simple interest on Le280 for 5 years at $2\frac{1}{2}\%$ per annum?

   A. Le245
   B. Le70
   C. Le35
   D. Le7
   E. Le3.50

5. What is the size of $\hat{p}$ in the diagram?

   A. 80°
   B. 70°
   C. 20°
   D. 160°
   E. 110°

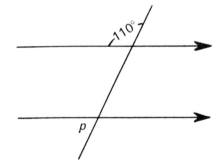

6. If $0 \cdot 000689 = 6 \cdot 89 \times 10^n$, what is the value of n?

   A. -4
   B. -3
   C. $+\frac{2}{3}$
   D. $+3$
   E. $+4$

7. In the diagram, △JKL ≡ △MNO, |JK| = 4", |JL| = 5", L̂ = M̂ and |KL| = |MO|. What is |MN|?

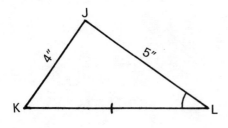

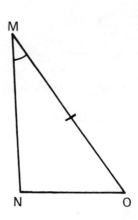

A. 3"

B. 4"

C. 4½"

D. 5"

E. More information is needed.

8. To pay his school fees for the coming term, a boy gets a job which pays him 4/6d. for a morning's work. If his fees are £3 16s. 6d., how many mornings must he work?

A. 15

B. 17

C. 18

D. 20

E. 21

9. In the right-angled triangle shown, |AB| = 15 in. and AB̂C = 36°. What is |AC| in inches?

A. $\dfrac{15}{\sin 36°}$

B. 15 sin 36°

C. $\dfrac{\sin 36°}{15}$

D. 15 cos 36°

E. 15 sin 54°

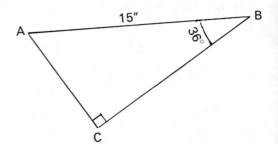

10. When a water tank is $x\%$ full it contains y gallons. What is the capacity of the tank?

   A. $x + y$ gal.

   B. $\dfrac{100x}{y}$ gal.

   C. $\dfrac{100y}{x}$ gal.

   D. $100xy$ gal.

   E. $\dfrac{y}{100x}$ gal.

11. The average of three numbers is 83·4. The average of two of the numbers is 86·8. What is the third number?

   A. 76·6

   B. 77·6

   C. 84·8

   D. 85·1

   E. 86·8

12. What is the value of m if $3^m \times 4 = 324$?

   A. 81

   B. 17

   C. 27

   D. 4

   E. 3

13. Which of the following has $(x - 1)$ as a factor?

   A. $5x^2 - x - 6$

   B. $5x^2 + x - 6$

C. $5x^2 + 7x - 6$

D. $5x^2 - 13x - 6$

E. $5x^2 + 13x - 6$

14. What is the volume of a torch battery which is in the form of a cylinder of diameter $1\frac{1}{4}$ in. and height $3\frac{1}{2}$ in? (Answer to one decimal place, take π to be $\frac{22}{7}$.)

   A. 3·4 cub. in.

   B. 4·3 cub. in.

   C. 8·6 cub. in.

   D. 17·2 cub. in.

   E. 27·5 cub. in.

15. A man and his son are standing together in the morning sun. The man, who is 6 feet tall, casts a shadow which is 40 feet long. What is the length of the shadow cast by his son who is 3 feet 9 inches tall?

   A. 40 ft.

   B. 39 ft.

   C. $27\frac{1}{2}$ ft.

   D. 25 ft.

   E. 24 ft.

16. In the diagram, $\overline{TX}$ is a tangent to the circle XYZ. YZT is a straight line and $|ZX| = |ZY|$. If $X\hat{Y}Z = 37°$, what is the size of $X\hat{T}Z$?

   A. 37°

   B. 69°

   C. 74°

   D. 79°

   E. 106°

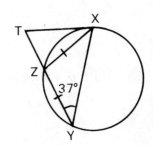

154

17. Express $\dfrac{3}{x-1} - \dfrac{4}{x+1}$ as a single fraction.

   A. $\dfrac{-x+7}{x^2-1}$

   B. $\dfrac{-x-1}{x^2-1}$

   C. $\dfrac{-x-7}{x^2-1}$

   D. $\dfrac{7}{x^2-1}$

   E. $\dfrac{-1+7x}{x^2-1}$

18. Given $2x + 3y = 6$ and $y - 3x = 1$, what is the value of $8x + y$?

   A. 8
   B. 7
   C. 6
   D. 5
   E. 4

19. Which of the following fractions is *not* equivalent to the base ten fraction $\tfrac{3}{7}{}_{ten}$?

   A. $\tfrac{3}{10}{}_{seven}$
   B. $\tfrac{3}{7}{}_{twelve}$
   C. $\tfrac{11}{111}{}_{two}$
   D. $\tfrac{10}{21}{}_{three}$
   E. $\tfrac{3}{13}{}_{five}$

20. In the diagram, $\overline{AB} \parallel \overline{CD}$. If the area of $\triangle DAE$ is 27 sq. cm. and the area of $\triangle DCE$ is 9 sq. cm., what is the area of $\triangle DBC$?

A. 18 sq. cm.

B. 36 sq. cm.

C. 3 sq. cm.

D. 25 sq. cm

E. More information is needed.

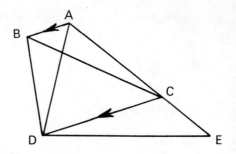

21. In the diagram, PQRS is a rectangle whose diagonals intersect at O. $\overline{OM} \parallel \overline{QR}$, $|PS| = 5''$ and $|MR| = 6''$. What is $|SQ|$?

    A. 7″
    B. 9″
    C. 11″
    D. 13″
    E. 17″

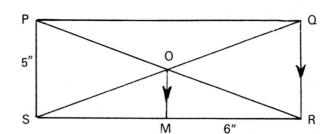

22. . . . and what is the area of quadrilateral OMRQ?

    A. 15 sq. in.
    B. 20 sq. in.
    C. 22½ sq. in.
    D. 30 sq. in.
    E. 45 sq. in.

23. What is 6·30452 to three significant figures?

    A. 6·30
    B. 6·304
    C. 6·305
    D. 6·31
    E. 6·35

24. If 1 kilogram is equivalent to 2·2 lb., how many grams are equivalent to 1 oz.?

   A. 28·4 gm.

   B. 32·5 gm.

   C. 35·2 gm.

   D. 62·5 gm.

   E. 137·5 gm.

25. In which of the following triangles can the equation,
    $|AB| = \dfrac{3 \sin 28°}{\sin 70°}$ inches, be used to calculate $|AB|$?

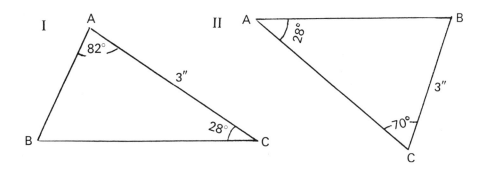

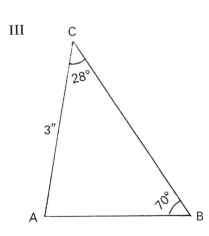

A. I and II

B. II and III

C. III only

D. I and III

E. I, II and III

26. Simplify $(x^6)^{1/2} \times (x^{1/2})^4$.

   A. x^5

   B. x^6

   C. x^{11}

   D. x

   E. None of these

27. In the diagram, $\overline{TA}$ is a tangent to the circle ABCD. If $\overline{CD}$ is a diameter of the circle and $D\hat{A}T = 35°$, what is the size of $A\hat{B}C$?

   A. 125°

   B. 145°

   C. 90°

   D. 110°

   E. 115°

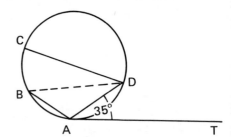

28. Four fishermen, A, B, C, D, sell their fish for £5 10s. The boat owner, A, takes 30% of this and the remainder is shared among B, C and D in the ratio 4:4:3 respectively. How much does D get?

   A. 40/-

   B. 33/-

   C. 30/-

   D. 28/-

   E. 21/-

29. What is the value of x if $\dfrac{1}{x} = \dfrac{1}{8} - \dfrac{1}{9}$?

   A. -72

   B. $-\frac{1}{72}$

   C. $\frac{1}{72}$

   D. -1

   E. 72

30. A man walks along a road at 4 m.p.h. for 2 hours. He then rests for 1 hour. Afterwards he gets a lift in a wagon which travels at 20 m.p.h. 1 hour later he completes his journey. What was his average speed for the journey?

   A. 12 m.p.h.

   B. 14 m.p.h.

   C. 8 m.p.h.

   D. 7 m.p.h.

   E. 6 m.p.h.

31. Given that the surface area of a sphere is four times the area of a plane circle of the same radius as the sphere, what is the surface area of a sphere of radius $1\frac{3}{4}$ inches? (Let π be $\frac{22}{7}$).

   A. $9\frac{5}{8}$ sq. in.

   B. $38\frac{1}{2}$ sq. in.

   C. 77 sq. in.

   D. 121 sq. in.

   E. 308 sq. in.

32. The energy, E, of a moving object is directly proportional to its mass, m, and the square of its velocity, v. Which of the following equations represents this information?

   A. $E = k(mv)^2$ (where k is a constant)

   B. $E = k(m + v^2)$ (where k is a constant)

   C. $E + E^2 = kmv$ (where k is a constant)

   D. $E = kmv^2$ (where k is a constant)

   E. $E = \dfrac{km}{v^2}$ (where k is a constant)

33. The altitudes AD, BE, CF of △ABC intersect at O.

   Which of the following statements is not true?

A. O is the centre of the in-circle of △ABC.
B. AFOE is a cyclic quadrilateral.
C. The area of △ABC is half the product of |AC| and |BE|.
D. D is not the midpoint of $\overline{BC}$.
E. $A\hat{B}E \neq E\hat{B}C$.

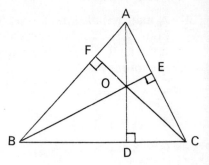

34. A trader bought some trinkets for 1/- each. Of these, 8 became so damaged that he was not able to sell them. He sold the remainder at 1/6d. each and made an overall profit of 33⅓%. How many trinkets did he buy to begin with?

   A. 96

   B. 80

   C. 72

   D. 64

   E. 36

35. If $x^2 + y^2 = 1$, what is x when $y = -\frac{3}{5}$?

   A. $\pm \frac{8}{25}$

   B. $\pm \frac{4}{25}$

   C. $\pm \frac{\sqrt{34}}{5}$

   D. $\pm \frac{16}{25}$

   E. $\pm \frac{4}{5}$

36. In the diagram, TPQ and TRS are tangents to the circle, centre O, at P and R. If $P\hat{R}S = 130°$, what is $P\hat{T}O$?

A. 30°
B. 40°
C. 50°
D. 60°
E. 80°

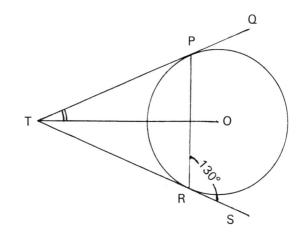

37. The sum of three consecutive numbers exceeds the square of the first number by 5. If n is the first number, which of the following is the equation obtained from this information?

    A. $3n = n^2 + 5$
    B. $3n = n^2 - 5$
    C. $3n + 3 = n^2 + 5$
    D. $3n + 3 = n^2 - 5$
    E. $3n + 3 = 5n^2$

38. What is |MN| in the diagram? (Note: $\cos 60° = \tfrac{1}{2}$)

    A. 8"
    B. 7½"
    C. 7"
    D. 6"
    E. 5½"

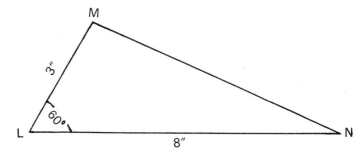

39. What is the length of the longest pole which could be put in a storeroom which is 12 ft. long, 9 ft. wide and 8 ft. high?

A. 8 ft.

B. 12 ft.

C. 15 ft.

D. 17 ft.

E. 21 ft.

40. 65% of the population of a town are not of school age. There are only enough schools to educate ⅘ of those who are of school age. What percentage of the townspeople can receive education?

   A. 80%

   B. 52%

   C. 40%

   D. 35%

   E. 28%

41. In the diagram, the cyclic quadrilateral is such that |KL| = |LM| = |MN|. If MK̂L = 55°, what is KM̂N?

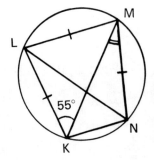

   A. 15°

   B. 20°

   C. 35°

   D. 45°

   E. 55°

42. If $(3p + q)^2 = a$, what is p in terms of a and q?

   A. $\dfrac{\pm\sqrt{a} - q}{3}$

   B. $\dfrac{a^2 - q}{3}$

162

C. $\pm \sqrt{\dfrac{a-q}{3}}$

D. $\dfrac{\frac{1}{2}a - q}{3}$

E. $\dfrac{\pm\sqrt{a-q}}{3}$

Use the given velocity triangle to answer the next two questions.

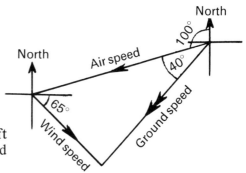

The velocity triangle for an aircraft flying in a wind. (All angles and description are on the diagram.)

43. Which of the following is the *course* of the aircraft?

   A. S. 40° W.

   B. S. 50° W.

   C. S. 80° W.

   D. N. 40° E.

   E. N. 50° E.

44. *From* which direction is the wind blowing?

   A. S. 35° E.

   B. N. 25° W.

   C. S. 25° E.

D. N. 35° W.

E. N. 65° W.

45. In the diagram, the radius of the outer circle is 2r and the small circles are each of radius r. Which of the following is the expression for the area of the shaded portion?

A. $4\pi r^2$

B. $3\pi r^2$

C. $2\pi r^2$

D. πr^2

E. $\tfrac{1}{2}\pi r^2$

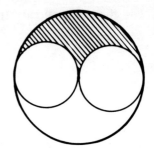

46. If log x + log x + log x = log y, what is the relationship between x and y?

A. $3x = y$

B. $\tfrac{1}{3}x = y$

C. $\sqrt[3]{x} = y$

D. $x^3 = y$

E. $x + 3 = y$

47. Evaluate $(4554)^2 - (4546)^2$.

A. 72,800

B. 72,000

C. 64,800

D. 64,000

E.   64

48. In the diagram, $A\hat{E}D = A\hat{C}B$. Which of the following statements is (are) true of the diagram?

   I △ABC ||| △ADE.

   II Points D, E, B, C are concyclic.

   III |AE| x |AB| = |AC| x |AD|.

   A. I, II and III
   B. II and III only
   C. I and II only
   D. I only
   E. II only

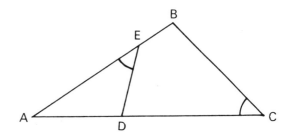

49. Two towns which are in latitude 40°S. have longitudes which differ by 90°. Taking the radius of the Earth to be R miles, what is the distance, in miles, between the towns along the parallel of latitude?

   A. $\frac{40}{360} 2\pi R \cos 90°$
   B. $\frac{90}{360} 2\pi R \cos 40°$
   C. $\frac{90}{360} R \cos 40°$
   D. $\frac{40}{360} 2\pi R$
   E. $\frac{90}{360} 2\pi R$

50. What is the value of x at the point of intersection of the graphs of the following equations?

$$5x - 2y - 14 = 0$$
$$3x - 7y + 9 = 0$$

   A. +3
   B. −3
   C. +4
   D. −4
   E. +5

Test 2
75 minutes 50 questions

1. Evaluate 0.0024×0.00000004.

    A. 9.6×10^{-10}

    B. 9.6×10^{-12}

    C. 9.6×10^{-11}

    D. 9.6×10^{11}

    E. 9.6×10^{10}

2. A town has 56,782 citizens. What is this number to three significant figures?

    A.   567

    B.   568

    C. 56,700

    D. 56,800

    E. 56,782·000

3. Factorize $4x^2 + x - 3$.

    A. $(2x + 3)(2x - 1)$

    B. $(2x - 3)(2x + 1)$

166

C. $(4x + 3)(x - 1)$

D. $(4x - 3)(x + 1)$

E. $(4x + 1)(x - 3)$

4. It is now x minutes to 1 p.m. 14 minutes ago it was x minutes past midday. What is the time now?

   A. 12:14 p.m.

   B. 12:23 p.m.

   C. 12:30 p.m.

   D. 12:37 p.m.

   E. 12:44 p.m.

5. In the diagram, O is the centre of the circle and $B\hat{C}A = 41°$. What is $B\hat{O}A$?

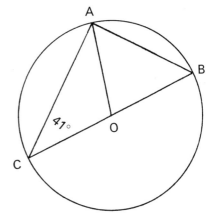

   A. 41°

   B. 49°

   C. 82°

   D. 98°

   E. 139°

6. ... and what is $B\hat{A}O$?

   A. 41°

   B. 49°

   C. 82°

   D. 98°

   E. 139°

167

7. What is the logarithm of $\sqrt[3]{0\cdot 0005957}$?

   A. $\bar{4}\cdot 7750$

   B. $\bar{4}\cdot 1986$

   C. $\bar{2}\cdot 9250$

   D. $\bar{1}\cdot 5917$

   E. $\bar{1}\cdot 2583$

8. What is the area of an equilateral triangle of length of side 5 in.? ($\sqrt{3} = 1\cdot 732$; answer to three significant figures.)

   A. 6·25 sq. in.

   B. 10·8 sq. in.

   C. 12·5 sq. in.

   D. 21·6 sq. in.

   E. 43·3 sq. in.

9. ϕ is an angle such that $\sin\phi = \tfrac{1}{2}$ and $\cos\phi = -\tfrac{\sqrt{3}}{2}$. What is the value of ϕ?

   A. 30°

   B. 60°

   C. 90°

   D. 120°

   E. 150°

10. The areas of two similar triangles are 8 sq. cm. and 18 sq. cm. The shortest side of the smaller triangle is 2 cm. in length. What is the length of the corresponding side of the larger triangle?

    A. 2¼ cm.

    B. 3 cm.

C. 3½ cm.

D. 4 cm.

E. 4½ cm.

11. The equation, $x^2 + 9x = 0$ has two roots. Which of the following is the greater root?

    A. 9

    B. −9

    C. 0

    D. 1

    E. 2

12. The numbers 5412 and 3643 were added and their sum was 11255. In what base were all these numbers?

    A. twelve

    B. ten

    C. nine

    D. eight

    E. seven

13. The length of an arc of a circle is one-sixth of the length of the circumference of the circle. What angle does the arc subtend at the centre of the circle?

    A. 30°

    B. 45°

    C. 60°

    D. 90°

    E. 180°

14. A ruler is in the form of a right prism whose cross-section, a trapezium, is shown in the diagram. If the greatest thickness of the ruler is one-tenth of an inch and its 'widths' are 1 inch and ½ inch, as shown, what is the volume of wood in the ruler if its length is 1 foot?

    A. 0·075 cub. in.
    B. 0·6 cub. in.
    C. 0·9 cub. in.
    D. 1·2 cub. in.
    E. 1·8 cub. in.

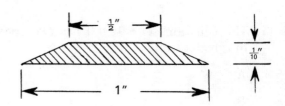

15. If $\log x = 3 \log 2$, what is the value of x?

    A. 6
    B. 5
    C. 9
    D. 8
    E. 32

16. In the diagram, O is the centre of the circle and $\overline{AB} \parallel \overline{CD}$. Which of the following statements about the diagram is (are) correct?

    I. $\triangle OBX \parallel \triangle DCX$
    II. $|AC| = |BD|$
    III. $\widehat{BXD} = 3 \times \widehat{BCD}$

    A. I only
    B. II only
    C. Both I and II
    D. Both I and III
    E. I, II and III are all correct.

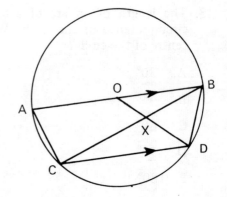

17. A square of side y inches has two of its adjacent sides increased by 1 in. and 8 in. respectively. The rectangle which results has an area of 1 square foot. Form an equation which could be solved for y.

   A. $y^2 + 9y - 136 = 0$

   B. $y^2 + 9y + 7 = 0$

   C. $2(y + 1) + 2(y + 8) = 48$

   D. $y^2 + 9y + 152 = 0$

   E. $y^2 = 144$

18. PQRS is a cyclic quadrilateral. If $P\hat{S}R = 78°$ and $P\hat{R}S = 36°$, what is $R\hat{Q}S$?

   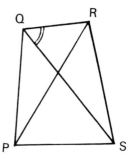

   A. 36°

   B. 42°

   C. 66°

   D. 51°

   E. 54°

19. What are the roots of the equation, $x^2 - 5x - 36 = 0$?

   A. +9, −4

   B. −9, +4

   C. −6, −6

   D. +6, −6

   E. −5, −36

20. △ABC is such that sides a, b, c are 6", 4", 3" in length respectively. Which of the following statements about △ABC is *not* true?

   A. △ABC is not a right angled triangle.

   B. $\hat{A}$ is the largest angle of the triangle.

171

C. $\hat{A}$ is obtuse.

D. $\hat{A}$ is twice the size of $\hat{C}$.

E. $\hat{B}$ and $\hat{C}$ must both be acute.

21. Use tables to evaluate $(\cos 40° - \cos 140°)$.

    A. 1·532

    B. 1·4088

    C. 0·1736

    D. 0·1232

    E. 0

22. A vertical telegraph pole, OY, is supported by two wires, YA and YB, where A, B and O are points on the horizontal ground such that $A\hat{O}B$ is a right angle. If the wire BY is 30 ft. long and A and B are 15 ft. and 25 ft. from O respectively, what is the height of the pole OY?

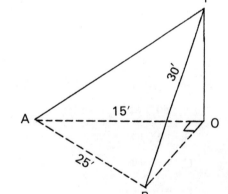

    A. 20 ft.

    B. 22·36 ft.

    C. 36·06 ft.

    D. 10 ft.

    E. 7·071 ft.

23. Express $\dfrac{1}{x+1} + \dfrac{2}{x^2-1}$ as a single fraction.

    A. $\dfrac{2}{x+1}$

    B. $\dfrac{x}{x-1}$

C. $\dfrac{x}{x^2 - 1}$

D. $\dfrac{2}{x^2 - 1}$

E. $\dfrac{1}{x - 1}$

24. △ABC stands between two parallel lines which are 6" apart. A lies on one parallel and $\overline{BC}$ on the other. If |BC| = 5", what is the area of △ABC?

   A. 15 sq. in.

   B. 30 sq. in.

   C. 11 sq. in.

   D. 7½ sq. in.

   E. 60 sq. in.

25. Consider the following statements:
   (a) A man runs a mile in 4 minutes.
   (b) A small boy runs 100 yards in 20 seconds.
   What is the ratio, (speed of the man) : (speed of the small boy)?

   A.   1:100

   B.   12:1

   C.   88:5

   D.   22:15

   E.   1056:5

26. Which of the following statements concerning circles and triangles is incorrect?
   A. A triangle has only one inscribed circle, but it has three e-scribed (or 'ex') circles.
   B. A triangle formed by joining the centre of a circle to the ends of any chord of the circle will be isosceles.
   C. The centre of a triangle's circumcircle is at the point of concurrence of the angle bisectors of the triangle.

D. The centre of the circumcircle of a right-angled triangle is at the mid-point of the hypotenuse of the triangle.

E. The circumcircle of an obtuse-angled triangle contains the triangle in its minor segment.

27. $\triangle ABC$ is right-angled at C. Consider the following statements and decide which is (are) true.

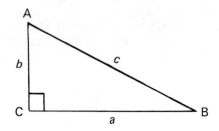

(i) $\cot \widehat{A} = \dfrac{b}{a}$

(ii) $\sec \widehat{B} = \dfrac{c}{a}$

(iii) $\csc \widehat{A} = \dfrac{c}{a}$

A. (i) only

B. Both (i) and (ii)

C. Both (ii) and (iii)

D. (iii) only

E. (i), (ii) and (iii) are all true.

28. In which of the diagrams below can the equation $\dfrac{a}{\sin 48°} = \dfrac{6}{\sin 73°}$ be used to calculate the length of a in inches?

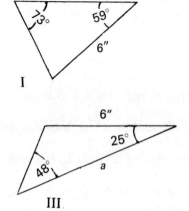

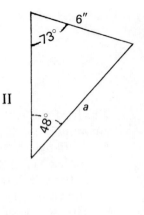

174

A. II only

B. I and II only

C. II and IV only

D. II and III only

E. I and IV only

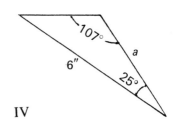

IV

29. The area of Ibadan's new Government Residential Area is ⅛ of a square mile. What area would the G.R.A. cover on a map whose scale is 1″ to 2 miles?

   A. $\frac{1}{32}$ sq. in.

   B. $\frac{1}{16}$ sq. in.

   C. $\frac{1}{8}$ sq. in.

   D. $\frac{1}{2}$ sq. in.

   E. 2 sq. in.

30. In the diagram, $\overline{AB}$ is a diameter and |AX| = |BX|. Given YX̂B = 38°, what is YÂX?

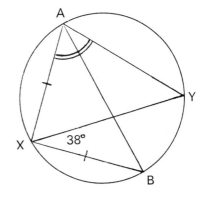

   A. 45°

   B. 83°

   C. 90°

   D. 97°

   E. 128°

31. A fan is revolving at 100 r.p.m. If the length of one of its blades is 1 ft. 9 in., through what distance does the outer tip of the blade move in ½ hour? (Take π to be $\frac{22}{7}$.)

   A. 33,000 ft.

B. 396,000 ft.

C. 11 ft.

D. 1,100 ft.

E. 28,875 ft.

32. It is said that a mosquito can detect the presence of a human being from 26 yards. Which of the following is the best description of the 'danger zone' for a man standing on open level ground?

    A. A circle, radius 26 yds

    B. A circle, radius 13 yds

    C. A cone, radius = height = 26 yds

    D. A hemisphere, radius 26 yds

    E. A cube whose edge is 26 yds in length.

33. Express the equation $\dfrac{2}{x+1} - \dfrac{3}{x} = \dfrac{3}{2}$ in a form which can be more easily solved.

    A. $3x^2 + 5x + 6 = 0$

    B. $3x^2 + 5x - 6 = 0$

    C. $3x^2 - 7x + 6 = 0$

    D. $3x^2 + x + 6 = 0$

    E. $3x^2 - x - 6 = 0$

34. What is the value of $(2x - y)(x^2 + y^2)$ when $x = 4$ and $y = -2$?

    A. 1

    B. 10

    C. 20

    D. 100

    E. 200

35. In the trapezium shown, $\overline{AB} \parallel \overline{DC}$, $|AD| = 15$ cm., $|AB| = 16$ cm., $|BC| = 17$ cm. and $A\hat{D}C$ is a right angle. What is $|DC|$?

A. 17 cm.

B. 24 cm.

C. 26 cm.

D. 31 cm.

E. 32 cm.

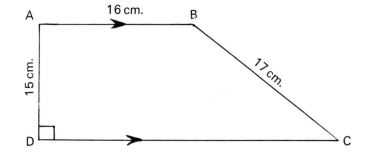

36. A fluorescent tube is 4 feet long and its diameter is 1·4 inches. Neglecting the thickness of the glass, what volume of gas is contained in the tube? (Use the approximation, $3\frac{1}{7}$, for π.)

   A. 73·92 cub. in.

   B. 10·56 cub. in.

   C. 211·2 cub. in.

   D. 6·16 cub. in.

   E. 295·68 cub. in.

37. What is the square root of the product of 0·0016 and 0·09?

   A. 0·12

   B. 0·012

   C. 0·0012

   D. 0·00072

   E. 0·00012

38. The combined weight of 6 men is 72 stones. If one of the men weighs 14 st. 2 lb., what is the average weight of the remaining men?

   A. 11 st. 8 lb.

   B. 12 st.

   C. 9 st. 9 lb.

   D. 17 st. 3 lb.

   E. 14 st. 5 lb.

39. 30 is to $2\frac{5}{8}$ as 1 is to . . .

   A. $78\frac{3}{4}$.

   B. $8\frac{3}{4}$.

   C. $11\frac{5}{7}$.

   D. $\frac{7}{80}$.

   E. 0·875.

40. In the diagram, O is the centre of the circle. ABE and DCF are straight lines, $A\widehat{O}B = 100°$, $E\widehat{B}C = 64°$ and $B\widehat{C}F = 78°$. What is $C\widehat{O}D$?

   A. 116°

   B. 102°

   C. 128°

   D. 100°

   E. 142°

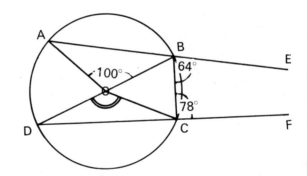

Use the given graph on page 179 of $y = 2x^2 - 5x + 1$ to answer questions 41, 42, 43.

41. What is the minimum value of $2x^2 - 5x + 1$?

   A. 2

   B. 0·4

   C. 0

   D. −0·4

   E. −2

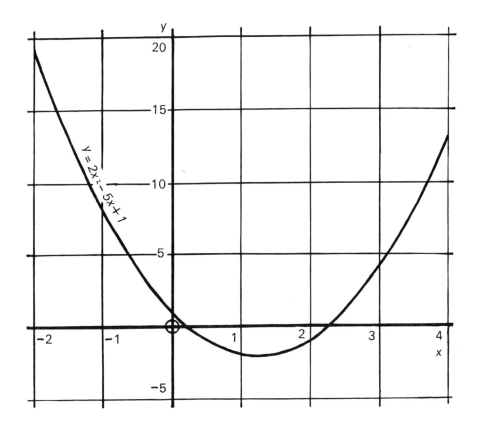

42. By making a suitable construction on the graph, find the gradient of the curve when $x = -1$.

  A. -9
  B. -2
  C. $+2$
  D. $+8$
  E. $+9$

43. Use the graph to solve the equation $2x^2 - 5x - 9 = 0$. The values of x which satisfy the equation are . . .

   A. $+0\cdot 2$ and $+2\cdot 2$.

   B. $-1\cdot 2$ and $+3\cdot 7$.

   C. $-1\cdot 1$ and $+3$.

   D. $-0\cdot 6$ and $+3\cdot 1$.

   E. 0 and $+2\cdot 5$.

44. In the diagram, $|AB| = |AC|$ and $|BD| = |BC|$. If $A\hat{B}D = x°$ and $D\hat{C}B = y°$, what is x in terms of y?

   A. $x = y$

   B. $x = 180 - y$

   C. $x = 180 - 2y$

   D. $x = 3y - 180$

   E. $x = 90 - y$

45. Which of the following diagrams does not contain a pair of congruent triangles?

   A. 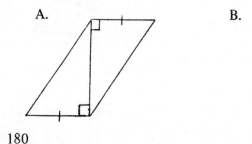   B.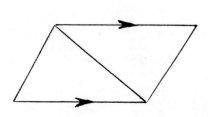

180

C.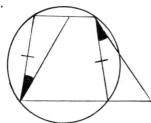

D.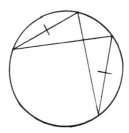

E.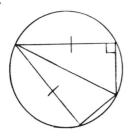

46. A pentagon has four of its sides equal in length. Two, and *only two*, of its interior angles are each equal to 63°. What is the size of each of the other three angles?

   A. 138°

   B. 126°

   C. 117°

   D. 103½°

   E. 99°

47. Three men start together to run round a running track. If one man can run the lap in 1 minute and the other two men take 1½ and 2 minutes respectively, how long will it be after the start until they are together again?

   A. 1½ min.

   B. 3 min.

C. 4½ min.

D. 6 min.

E. 13 min.

48. What is x if $2^x = \frac{1}{8}$?

   A. $\frac{1}{16}$

   B. $\frac{1}{4}$

   C. $\frac{1}{3}$

   D. -3

   E. -4

49. In the diagram, trapezium ABCD has $\overline{AB} \parallel \overline{DC}$. P is the point of intersection of the diagonals and $|PA| = |PB|$. In what logical order must statements 1, 2, 3, 4 be taken in proving that ABCD is a cyclic quadrilateral?

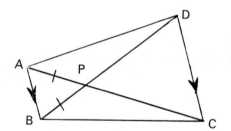

1. $B\hat{A}C = B\hat{D}C$ (both shown equal to $A\hat{B}P$)
2. $A\hat{B}P = P\hat{A}B$ (angles opp. equal sides of $\triangle APB$)
3. ABCD is a cyclic quad.   (angles in the same segment equal)
4. $A\hat{B}P = B\hat{D}C$ (alt. angles, $\overline{AB} \parallel \overline{DC}$)

   A. 1, 2, 3, 4
   B. 1, 2, 4, 3
   C. 2, 1, 4, 3
   D. 4, 1, 2, 3
   E. 4, 2, 1, 3

50. A sector of a circle has an angle of 30° and its area is 10π sq. cm. Which of the following is an expression for the length of the radius of the circle?

A. $\sqrt{120}$ cm.

B. $\sqrt{60}$ cm.

C. 60π cm.

D. $\sqrt{60\pi}$ cm.

E. $\dfrac{10\pi}{6}$ cm.

183

Test 3
75 minutes 50 questions

1. Express p shillings and q pence in Pounds (£).

   A. $£\dfrac{p + 12q}{240}$

   B. $£\dfrac{12p + q}{240}$

   C. $£\dfrac{p + q}{20}$

   D. $£\left(\dfrac{p}{20} + \dfrac{q}{12}\right)$

   E. $£\dfrac{20p + 12q}{240}$

2. The volume of a wooden ruler is 18 cc. and its density is 0·4 g./cc. What is the weight of the ruler?

   A. 45 g.

   B. $22\tfrac{2}{3}$ g.

   C. 7·2 g.

   D. 4·5 g.

   E. 0·72 g.

3. Express 4½ as a decimal fraction of 40.

   A. 8·889

   B. 0·45

   C. 1·8

   D. 0·225

   E. 0·1125

4. The equation $\frac{1}{f} = \frac{1}{v} + \frac{1}{u}$ shows a connexion between f, v and u. If $v = 9$ and $u = 36$, what is the value of f?

   A. 12

   B. $7\tfrac{1}{5}$

   C. 4

   D. 1

   E. $\tfrac{5}{36}$

5. Two pairs of canvas shoes cost 10/10d. per pair. Six other pairs of canvas shoes cost £3 11s. all together. If someone buys all the shoes, what is the average cost per pair?

   A. 11/1d.

   B. 11/4d.

   C. 11/7d.

   D. 11/10d.

   E. 12/1d.

6. What is the square root of 45,986? (To three significant figures).

   A. 214

   B. 67·8

   C. 21·4

D. 678

E. 2140

7. A bank will give £N1 in exchange for $2.76. How many pounds would be given in exchange for $23?

   A. £N63.48

   B. £N25

   C. £N12

   D. £N8½

   E. £N8⅓

8. If $x - 3y = -4$ and $3x + 2y = 8$, what is the value of $2x + 5y$?

   A. 1

   B. 4

   C. 12

   D. −12

   E. 20

9. One of the five fractions below is equal to $(\frac{2}{3})^{-4}$. Which one?

   A. $-\frac{8}{12}$

   B. $-\frac{16}{81}$

   C. $\frac{16}{81}$

   D. $\frac{81}{16}$

   E. $-\frac{81}{16}$

10. Ahmed, Bala and Chado were asked to write down one factor of the expression, $2x^2 + x - 3$. Here is what they wrote:
    Ahmed, $(2x - 3)$; Bala, $(x - 1)$; Chado, $(2x + 3)$.
    Which boy(s) was (were) correct?

    A. Ahmed only

    B. Bala only

C. Chado only

D. Ahmed and Bala

E. Bala and Chado

11. In the diagram, which of the lines AOB, COD, EOF, GOH are straight?

   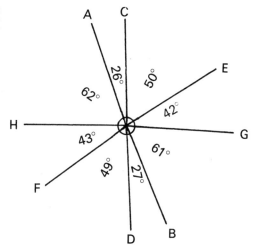

   A. AOB and FOE

   B. COD and GOH

   C. Only GOH

   D. None of them is straight.

   E. They are all straight.

12. Which of the following statements is *not* incomplete?

   A. One mile equals 1760.

   B. A tangent to a circle makes an angle of 90°.

   C. In a circle, equal arcs subtend equal angles.

   D. An equilateral triangle is also equiangular.

   E. Two triangles are equal in area if they stand on the same base.

13. Take $3a + 2y - 5z$ from $a - 3y + 5z$.

   A. $2a + 5y - 10z$

   B. $-2a + y$

   C. $-2a - 5y + 10z$

   D. $2a - 5y$

   E. $-2a + y - 10z$

14. The equation, $3x^2 + 2x = 0$ has two roots. If these roots are multiplied, one by the other, what will be the product?

   A. 0

   B. $\frac{2}{3}$

   C. 6

   D. $-\frac{2}{3}$

   E. $-1\frac{1}{2}$

15. A tube of 'Smarties', containing 35 different coloured chocolate buttons, is to be shared among four children. Abioseh takes all the red ones and the remainder is shared among the other three children in the ratio 4:3:2, of which the smallest share is 6 'Smarties'. How many red 'Smarties' were there?

   A. 6

   B. 8

   C. 9

   D. 12

   E. 24

16. In △ABC, |AB| = 5 in., |BC| = 9 in. and |CA| = 6 in. What is the cosine of BÂC?

   A. $\frac{1}{3}$

   B. $\frac{2}{3}$

   C. $-\frac{7}{9}$

   D. $-\frac{1}{3}$

   E. $\frac{5}{9}$

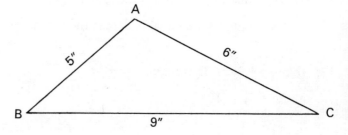

17. The value of a car depreciates by 25% of its cost price in the first year after purchase. In its second year after purchase its value depreciates by 20% of what its value was at the beginning of that year. What

percentage of its original cost price is the car worth at the end of the second year?

A. 50%

B. 55%

C. 60%

D. 65%

E. 70%

18. If $2n$ is an even number, which of the following must be an odd number?

A. n

B. $n - 1$

C. $n + 1$

D. $3n$

E. We cannot say unless the value of n is given.

19. In the diagram, O is the top of a flagpole and points R, O and S all lie in the same vertical plane. The angle of depression of S from the top of the flagpole is 48°, and the angle of elevation of O from R is 63°. What is RÔS?

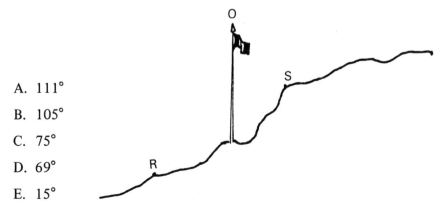

A. 111°

B. 105°

C. 75°

D. 69°

E. 15°

20.

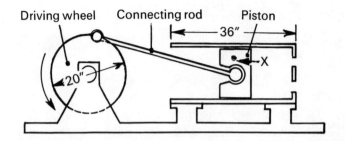

The diagram represents the main working parts of a pump mechanism. The driving wheel, diameter 20", drives the piston by means of the connecting rod. What is the locus of the point X on the piston?

A. A circle, radius 10"

B. A straight line, length 36"

C. A straight line, length 20"

D. A straight line, length 10"

E. An arc of a circle

21. The speed, V, of a car is directly proportional to the amount of petrol, P, it uses and the square root of its engine capacity, c, and is inversely proportional to the weight, W, of its occupants. What is V in terms of P, c and W (and any constants that may be required)?

A. $V = kP\sqrt{c}W$ (where k is a constant)

B. $V = kPcW$ (where k is a constant)

C. $V = kP\sqrt{c}/W$ (where k is a constant)

D. $V = aP\sqrt{c} + (b/W)$ (where a and b are constants)

E. $V = aP + b\sqrt{c} + (k/W)$ (where a, b and k are constants)

22. How many sides has a regular polygon, if each interior angle is 165°?

A. 72

B. 36

C. 30

D. 24

E. 15

23. What is the sum of te_{twelve} and et_{twelve}?

    A. teet
    B. 231
    C. 273
    D. 1t9
    E. 42

    (Note, t and e are used here as symbols for the digits $t$en and $e$leven respectively)

24. A farmer has c cows and s sheep. If he had four more cows, he would have twice as many cows as sheep. Which of the following equations is a correct interpretation of this information?

    A. $2c + 4 = s$
    B. $c + 4 = 2s$
    C. $c + 4 = \tfrac{1}{2}s$
    D. $2c + 4 = \tfrac{1}{2}s$
    E. $2c + 4 = 2s$

25. In the diagram, APQB ∥ CRD and $P\widehat{R}Q$ is a right angle. Which of the following statements is incorrect?

    A. $x + y = 90°$
    B. $a - c = 90°$
    C. $x + b = 90°$
    D. $y + b = 90°$
    E. $b + c = 90°$

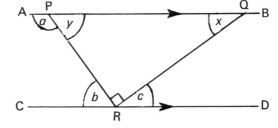

26. Given, $a = ay + x$, what is a in terms of x and y only?

   A. $\dfrac{x}{1-y}$

   B. $\dfrac{x}{1+y}$

   C. $\dfrac{y-1}{x}$

   D. $\dfrac{x}{y}$

   E. $\dfrac{x-y}{x}$

27. What is the value of $2x^2 - 5y - z^2$ when $x = -2, y = -4$ and $z = -6$?

   A. -48

   B. -66

   C. -12

   D. $+66$

   E. -8

28. A solid brass cube is melted down and recast as a solid cone of height h and base radius r. If the height of the cube was also h, what is r expressed in terms of h?

   A. $r = h$

   B. $r = h\sqrt{\dfrac{3}{\pi}}$

   C. $r = \pi h$

   D. $r = \sqrt{\dfrac{3h}{\pi}}$

   E. $r = \dfrac{1}{h}\sqrt{\dfrac{\pi}{3}}$

29. In the diagram, A is a point on $\overline{BD}$ and $\overline{AE} \parallel \overline{BC}$. If $|CA| = |CB|$ and $D\hat{A}E = 36°$, what is the size of $D\hat{A}C$?

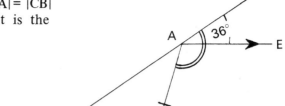

A. 144°

B. 118°

C. 108°

D. 76°

E. 72°

30. Given a line AB, how would you locate a point, P, such that P is always in a position which makes $A\hat{P}B$ a right angle? P would be ...

A. any point on the perpendicular bisector of AB.

B. any point on a line parallel to AB.

C. any point on the circumference of the circle whose diameter is AB.

D. any point on the circumference of a circle which has AB as radius.

E. the midpoint of AB.

31. The straight line graphs of $y = -2x + 1$ and $y = 3x + 11$ intersect at the point F. What is the value of x at F?

A. 3

B. 2

C. 1

D. −1

E. −2

32. In $\triangle ABC$, $\cos \hat{A} = -0.8282$. Which of the following deductions is correct?

A. $\hat{A} = 34° \; 5'$

B. $\overline{AB}$ is the shortest side of the triangle.

C. $\widehat{B}$ is obtuse.

D. $\overline{BC}$ is the longest side of the triangle.

E. $\widehat{A} = 80°\ 7'$

33. In the diagram, $\overline{AB}$ is the diameter of the semi-circle ACB. ACD is a straight line and $\overline{DB}$ is a tangent to the semi-circle at B. If the radius of the semi-circle is 5 cm. and $|BC| = 6$ cm., what is $|CD|$?

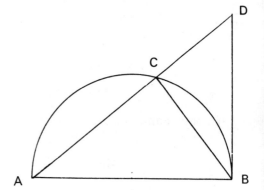

A. $4\frac{1}{2}$ cm.

B. $6\frac{1}{4}$ cm.

C. $7\frac{1}{2}$ cm.

D. 8 cm.

E. $12\frac{1}{2}$ cm.

34. If m is any digit, which of the following statements must be true in the evaluation of $\sqrt{0 \cdot m}$?

A. The result will be less than $0 \cdot m$.

B. The result will be positive.

C. The result will be negative.

D. $0 \cdot m$ cannot be negative.

E. The first significant figure of the result could be 1 or 2.

35. $\triangle PQR$ is such that $|PQ| = |PR| = 10''$ and $QR = 12$. What is the area of the triangle?

A. 48 sq. in.
B. 50 sq. in.
C. 60 sq. in.
D. 96 sq. in.
E. 120 sq. in.

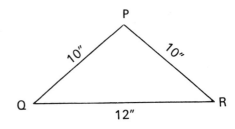

36. Find the value of $\frac{a}{b}$ if $\frac{3b - 2a}{a + 2b} = 3$.

    A. $\frac{3}{5}$
    B. $-\frac{3}{5}$
    C. $\frac{5}{3}$
    D. $-\frac{5}{3}$
    E. $-\frac{1}{3}$

37. In the figure, $\overline{AE}$ and $\overline{BD}$ are two of the altitudes of $\triangle ABC$. If $|BC| = 6"$, $|AE| = 4"$ and $|AC| = 8"$, what is $|BD|$?

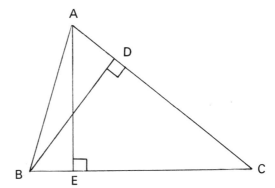

    A. 3"
    B. 4"
    C. 5⅓"
    D. 6"
    E. 8"

38. In each of the given figures there are four points A, B, C, D. By considering the information in each figure, state the figures in which a circle could be drawn through the points A, B, C, D.

195

A. I and II only

B. II and IV only

C. I and IV only

D. IV only

E. II and III only

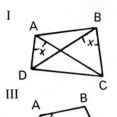
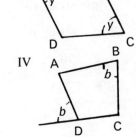

39. A quadrant of a circle of radius r is cut away from each corner of a square. If the length of a side of the square is $2r$, what is the length of the perimeter of the remainder?

A. $8r - 2\pi r$

B. $r^2(4 - \pi)$

C. $8r - \pi r^2$

D. $4r^2 - 2\pi r$

E. $2\pi r$

40. The following is an arithmetic addition using numbers in different bases:

$$442_{five} + 224_{five} = X_{ten}$$

If the addition is correct, what is X?

A. 186

B. 333

C. 666

D. 1221

E. 2442

41. In the diagram, O is the centre of the circle. If $\hat{x} = 100°$, what is $\hat{y}$?

A. 80°
B. 100°
C. 130°
D. 160°
E. 200°

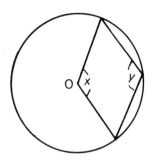

42. An open rectangular box has external dimensions: *a* ft. long by *b* in. wide by *c* in. high. If the box is 1 in. thick, what is the volume of material used in making the box?

   A. $12abc - (12a - 2)(b - 2)(c - 2)$ cub. in.

   B. $12abc - (12a - 2)(b - 2)(c - 1)$ cub. in.

   C. $abc - (a - 2)(b - 2)(c - 1)$ cub. in.

   D. $24ac + 12ab + 2bc$ cub. in.

   E. $(12a + 2)(b + 2)(c + 1) - 12abc$ cub. in.

43. In the diagram, the circle is inscribed in △ABC, and two of the angles are given. What is the size of BÂC?

   A. 42°
   B. 49°
   C. 56°
   D. 62°
   E. 69°

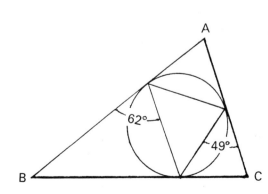

44. The figure shows a triangle which is right angled at A. An altitude has been drawn from A to the opposite side and all the line segments have been given letters to denote their length. Which of the following statements is not correct?

A. $\dfrac{a}{p} = \dfrac{q}{r}$

B. $\dfrac{a}{p} = \dfrac{p}{a+b}$

C. $q^2 = ab$

D. $\dfrac{p}{q} = \dfrac{r}{b}$

E. $\dfrac{r}{p} = \dfrac{q}{b}$

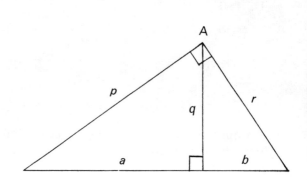

45. $\overline{XY}$ is a diameter of circle XPY, centre O. $\overline{XL}$ and $\overline{YM}$ are perpendiculars from X and Y to the tangent at P. In what logical order must the statements 1, 2, 3, 4, 5 be placed in proving $|PL| = |PM|$?

Statements: 1 $\overline{XL} \parallel \overline{OP} \parallel \overline{YM}$ ...   (corresp. angles)
2 $M\widehat{P}O = 1$ right angle ...   (tan. and rad.)
3 $|PL| = |PM|$ ...   (equal intercept th.)
4 $|OX| = |OY|$ ...   (radii)
5 Join OP.

A. 5, 1, 2, 4, 3

B. 5, 2, 1, 4, 3

C. 5, 1, 2, 3, 4

D. 5, 4, 1, 2, 3

E. 1, 4, 5, 2, 3

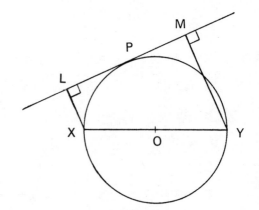

46.

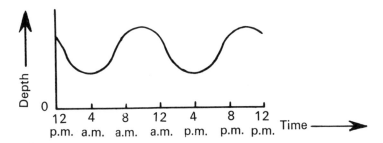

The graph shows how the depth of water in a tidal river varies throughout a day..Which of the following statements is not true?

A. High tides occurred at 10 a.m. and 10 p.m.

B. There was a low tide at 4 a.m.

C. There was no water in the river at 4 p.m.

D. There were two high tides and low tides during the day.

E. At 8 a.m. the next low tide would be in 8 hours time.

47. 'Pickem Pools, Ltd' spends its takings in the following way. 40% of the takings goes on tax, costs and profit. 60% of the remainder is given to the 1st prize winner. The money left is shared in the ratio 6:4 between the 2nd and 3rd prize winners. What percentage of the original intake does the third prize winner get?

A. 9·6%

B. 10%

C. 14·4%

D. 24%

E. 36%

48. Towns P and Q, both on latitude 23°S., are situated so that P is 18° west of Q. If R is the radius of the Earth, what is their distance apart along the parallel of latitude?

A. $\frac{18}{360} 2\pi R$

B. $\frac{23}{360} 2\pi R \cos 18°$

199

C. $\frac{1}{360} 2\pi R \cos 23°$

D. $\frac{23}{360} R \cos 18°$

E. $\frac{18}{360} 2R \cos 23°$

49. In the diagram, $\overline{AC}$ and $\overline{BD}$ are the diagonals of cyclic quadrilateral ABCD.

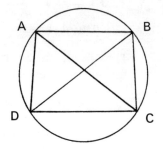

Consider the statements I, II and III and decide which is (are) true.

   I. If $|AB| = |DC|$, ABCD will be a trapezium.
  II. If $|AC| = |BD|$, either $\overline{AB} \parallel \overline{CD}$, or $\overline{AD} \parallel \overline{BC}$.
 III. If $D\widehat{A}B = A\widehat{B}C = 90°$, ABCD will be a square.

A. I only

B. II only

C. III only

D. Both I and II

E. Both I and III

50. The diagram shows a view of a square based pyramid. The vertex V is placed above A (of base ABCD) so that $V\widehat{A}B$ is a right angle. $|VA| = 2\frac{1}{2}"$ and $|AB| = 2"$.

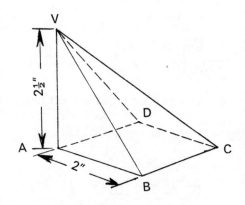

Which of the drawings A, B, C, D, E is the elevation of the solid in the vertical plane parallel to $\overline{BC}$?

A.
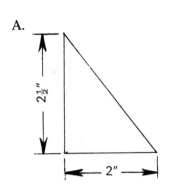

B.
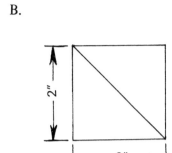

C.

D.
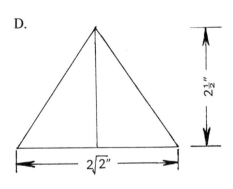

E.
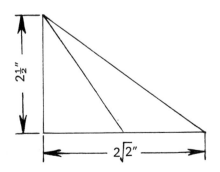

Test 4
75 minutes 50 questions

1. What is 11·80563 to three decimal places?

    A. 11·8

    B. 11·805

    C. 0·806

    D. 11·806

    E. 11·810

2. Solve the equation, $\frac{1}{2}x - \frac{2}{3} = 1$.

    A. $x = 10$

    B. $x = 3\frac{2}{3}$

    C. $x = 3\frac{1}{3}$

    D. $x = \frac{5}{6}$

    E. $x = \frac{2}{3}$

3. In the diagram, two of the exterior angles of $\triangle ABC$ are shown to be 131° and 142°. What is the size of $A\widehat{C}B$?

A. 83°
B. 87°
C. 90°
D. 93°
E. 100°

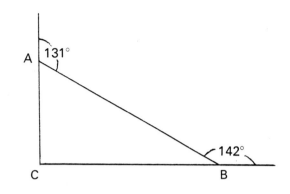

4. What is the average height of four boys whose heights are 4 ft. 4 in., 4 ft. 9 in., 4 ft. 11 in. and 5 ft. 8 in.?

   A. 4 ft. 11 in.
   B. 4 ft. 10 in.
   C. 4 ft. 9 in.
   D. 4 ft. 7 in.
   E. 19 ft. 8 in.

5. Evaluate $\dfrac{2x - ay}{y}$ if $a = 3$ when $x = 7$ and $y = 4$.

   A. 17
   B. 11
   C. $6\frac{1}{2}$
   D. $1\frac{3}{4}$
   E. $\frac{1}{2}$

6. What is the value of $4 \times \sqrt{25 \times 169}$?

   A. 65
   B. 260
   C. 1040
   D. 1300
   E. 16900

203

7. In △ACB, |AC| = |CB|, AĈB = 50° and the sides BA and CA are produced to E and F respectively. What is the size of FÂE?

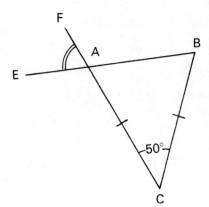

A. 25°
B. 50°
C. 65°
D. 115°
E. 130°

8. Factorise $6x^2 + 23x - 4$.

A. $(6x - 1)(x + 4)$
B. $(6x + 1)(x - 4)$
C. $(3x - 2)(2x + 2)$
D. $(x - 1)(6x + 4)$
E. $(x + 1)(6x - 4)$

9. In an examination a boy who scores 154 marks gets 44%. What is the total for the examination?

A. 29 marks
B. 100 marks
C. 198 marks
D. 350 marks
E. 400 marks

10. Points A, B, C, D are on the circumference of a circle so that AB̂C = 117° and AĈD = 26°. What is the size of CÂD?

A. 26°
B. 37°
C. 63°
D. 89°
E. 91°

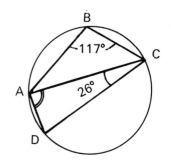

11. It is required to calculate |AB| in the given △ABC. Which of the following should be used?

    A. The sine rule
    B. The cosine rule
    C. Pythagoras' theorem
    D. Congruent triangles
    E. The extension to Pythagoras' theorem

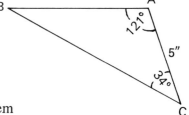

12. Bala's wage is twice as much as Alpha's, and Chike's wage is the average of the other two wages. What is the ratio of these wages, expressed in the order, Alpha's wage: Bala's wage: Chike's wage?

    A. 1:2:3
    B. 1:2:2
    C. 2:4:5
    D. 4:2:3
    E. 2:4:3

13. Simplify $\dfrac{3x - 12}{x^2 - 8x + 16}$.

    A. $1\frac{1}{2}$
    B. 3

C. $\dfrac{3}{x+4}$

D. $\dfrac{3}{x-4}$

E. $\dfrac{3(x-1)}{x^2-2}$

14. WXYZ is a parallelogram and $\overline{WP}$ is the perpendicular from W to $\overline{XY}$. If the area of the parallelogram is 60 sq. in. and |WZ| = 10 in., what is |WP|?

    A. 6 in.
    B. 5 in.
    C. 12 in.
    D. 300 in.
    E. More information is needed.

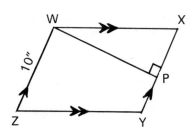

15. A man has £a, b shillings and c pence. What is this amount in pence?

    A. $\dfrac{a}{240} + \dfrac{b}{20} + c$

    B. $240a + 12b + c$

    C. $a + \dfrac{b}{20} + 240c$

    D. $240a + 20b + c$

    E. c

Use the following graph to answer questions 16, 17, 18.
The graph shows how Isah left home on his bicycle, stopped to mend a puncture, and then returned home.

16. How long did it take Isah to mend the puncture?

    A. ¾ hour

B. ½ hour

C. ¼ hour

D. 1 hour

E. 4 hours

17. Neglecting the time it took him to mend the puncture, what was Isah's average *cycling* speed for the outing?

    A. 4 m.p.h.

    B. 5 m.p.h.

    C. 6⅔ m.p.h.

    D. 7½ m.p.h.

    E. 10 m.p.h.

18. What was Isah's average speed for the *whole* outing?

    A. 4 m.p.h.

    B. 5 m.p.h.

    C. 6⅔ m.p.h.

    D. 7½ m.p.h.

    E. 10 m.p.h.

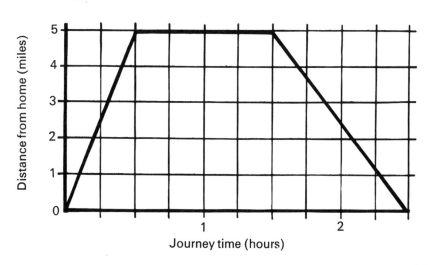

19. Planks of mahogany wood are sold in the market for 13/- each. If the planks are 10 feet long by 1 foot wide by 1 inch thick, what is the cost of the wood per cubic foot?

   A.   1·3 shillings

   B.   10/10d.

   C.   15·6 shillings

   D.   31·2 shillings

   E.   130 shillings

20. The gear on the pedals of a bicycle has 56 teeth and the gear on the back wheel has 14 teeth. How many times does the back wheel rotate for one revolution of the pedals?

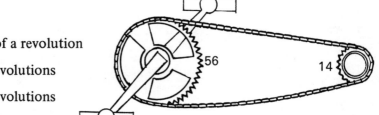

   A.   ¼ of a revolution

   B.   2 revolutions

   C.   4 revolutions

   D.   14 revolutions

   E.   We need to know the length of the chain.

21. If $x^2 + 5x + k$ is a perfect square, what must be the value of k?

   A.   25

   B. $\frac{25}{4}$

   C.   5

   D. $\frac{5}{2}$

   E. $\frac{25}{16}$

22. What is the sum of the *binary* fractions, $\frac{1}{11}$ and $\frac{1}{10}$?

   A. $\frac{21}{110}$

   B. $\frac{2}{21}$

C. $\frac{1}{110}$

D. $\frac{101}{110}$

E. $\frac{11}{110}$

23. If the ratio, $x:y = 2:5$, what is the ratio $3x + 2y:3x + 4y$?

    A. 8:13
    B. 1:4
    C. 19:23
    D. 2:5
    E. 1:10

24. If $3x + 2y:3x + 4y = 3:5$, what is the ratio $x:y$?

    A. 2:5
    B. 1:3
    C. 3:5
    D. 5:3
    E. 19:29

25. In the diagram, $\overline{TA}$, $\overline{TB}$ are tangents to the circle, centre O. Which of the following statements is not necessarily correct?

    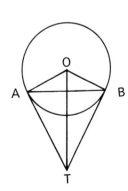

    A. △ AOB is isosceles.
    B. △ TAB is equilateral.
    C. $\overline{OT}$ perpendicularly bisects $\overline{AB}$.
    D. △ s AOT and BOT are congruent.
    E. TAOB is a cyclic quadrilateral.

26. A monthly electricity bill is made up of a fixed charge of 15/6d., after which the cost of the number of units of electricity is added. If the units are priced at 5d. each, what is the bill for someone who uses 111 units in a month?

   A. 15s. 6d

   B. £1 7s. 9d.

   C. £2 3s. 3d.

   D. £2 6s. 3d.

   E. £3 1s. 9d.

27. If $2^x = \frac{1}{32}$, what is the value of x?

   A. $\frac{1}{64}$

   B. $\frac{1}{16}$

   C. -5

   D. $-\frac{1}{5}$

   E. 5

28. What is the volume of the metal in a Nigerian penny if the inner and outer radii are r and R respectively and its thickness is t?

   A. $\pi(R^2 - r^2)t$

   B. $\pi(R - r)^2 t$

   C. $2\pi(R - r)t$

   D. Rrt

   E. $\pi(R - r)t^2$

29. The Earth rotates once about its own axis every day. How long does it take to rotate through $1°$?

   A. 1 min.

   B. $3\frac{3}{4}$ min.

   C. 4 min.

D. 7½ min.

E. 15 min.

30. A market woman asks 1/3d. for a measure of rice. After some bargaining, she sells 5 measures for 5/6d. By what percentage did she reduce her starting price?

   A. 3%
   B. 6%
   C. 9%
   D. 12%
   E. $13\frac{7}{11}\%$

31. What are the solutions of the equations, $x^2 - y^2 = -12$ and $x + y = 4$?

   A. $x = -3\frac{1}{2}, y = -\frac{1}{2}$
   B. $x = -\frac{1}{2}, y = -3\frac{1}{2}$
   C. $x = 4\frac{1}{2}, y = -\frac{1}{2}$
   D. $x = 3\frac{1}{2}, y = \frac{1}{2}$
   E. $x = \frac{1}{2}, y = 3\frac{1}{2}$

32. Quadrilateral ABCD is such that $|AB| = |AC| = |AD|$. If $B\widehat{A}D = 70°$, what is the size of $B\widehat{C}D$?

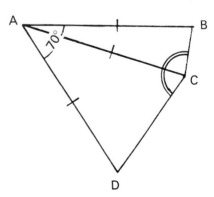

   A. 145°
   B. 140°
   C. 135°
   D. 130°
   E. More information is needed.

33. The post office savings bank pays simple interest at 2½% per annum. What was the interest on N₵20 which was deposited on 1 January 1967 and withdrawn on 1 July 1968?

   A.   50 Np
   B.   75 Np
   C.   80 Np
   D.   N₵1.25
   E.   N₵2.50

34. A right pyramid, VPQRS, has its vertex, V, 5 inches above its rectangular base PQRS. If |PQ| = 8" and |QR| = 6", what is |VP|?

   A. 5"
   B. 5·59"
   C. 6"
   D. 7·07"
   E. 8"

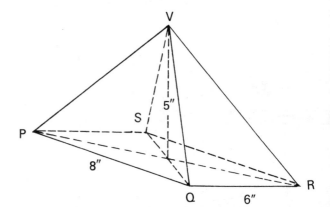

35. A wheel of radius 2 ft. 4 in. makes 132 revolutions while travelling along a road. How far does the wheel travel? (Approximate π to 22/7).

   A.   9 ft.
   B.   132 ft.
   C.   1936 ft.
   D.   2265 ft. 8 in.
   E.   2904 ft.

36. A man cycles a distance of $3a$ miles at v m.p.h. and then walks a distance of a miles at $(v - 7)$ m.p.h. What was the total number of hours he spent travelling?

A. $\dfrac{3a}{v} + \dfrac{a}{v-7}$

B. $\dfrac{v}{3a} + \dfrac{v-7}{a}$

C. $\dfrac{4a}{2v-7}$

D. $\dfrac{2v-7}{4a}$

E. $3av + a(v-7)$

37. From the graph, $y = x^2 - x + 5$, it is required to solve the equation $2x^2 = 2x - 7$. The solutions are found where the curve meets the line ...

   A. $y = -1\tfrac{1}{2}$

   B. $y = 1\tfrac{1}{2}$

   C. $y = -2$

   D. $y = 3\tfrac{1}{2}$

   E. $y = 7$

38. In the diagram, straight line ATB is a tangent to the circle TCD, where $|TC| = |TD|$. In what order must the following statements be placed in proving that $\overline{AB} \parallel \overline{CD}$?

   I. $T\hat{C}D = T\hat{D}C$ (angles opp. equal sides, $\triangle CTD$)
   II. $\overline{AB} \parallel \overline{CD}$ (alternate angles equal)
   III. $A\hat{T}C = T\hat{C}D$ (both equal to $T\hat{D}C$)
   IV. $A\hat{T}C = T\hat{D}C$ (alt. segment th.)

   A. I, II, III, IV

   B. I, III, IV, II

   C. I, IV, III, II

   D. I, IV, II, III

   E. IV, III, II, I

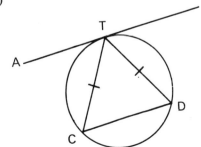

213

39. 5 lb. of sugar cost x cents; 4 lb. of another brand cost y cents. The difference in cost between the brands is 1c. per lb., the sugar which sells at x cents for 5 lb. being more expensive. Which of the following equations represents this information?

   A. $5x - 4y = 1$

   B. $4x - 5y = 1$

   C. $5x - 4y = 20$

   D. $4x - 5y = 20$

   E. None of these

40. A straight line, MN, 6" long, is drawn on a large sheet of paper. How many points can be drawn on the paper such that each point is 2" from MN *and* such that MN subtends an angle of 90° at each point?

   A. 8

   B. 4

   C. 2

   D. 1

   E. 0

41. A faulty speedometer of a car shows 26 m.p.h. when the car is really doing 34 m.p.h. If the error in the reading varies as the true speed of the car, what is the actual speed of the car when the speedometer shows 65 m.p.h.?

   A. 45 m.p.h.

   B. 57 m.p.h.

   C. 65 m.p.h.

   D. 73 m.p.h.

   E. 85 m.p.h.

42. Which of the following statements is true about a parallelogram, but not true about a rectangle?

   A. Its opposite sides are equal in length.

B. Its opposite angles are equal.

C. Its diagonals are not equal in length.

D. Its opposite sides are parallel.

E. Its diagonals bisect each other.

43. A bird, sitting on the branch of a tree, drops vertically down to the ground through 8 ft. It then walks 6 ft., changes direction through 90° and walks a further 24 ft. This is shown by the path ABCD in the diagram:

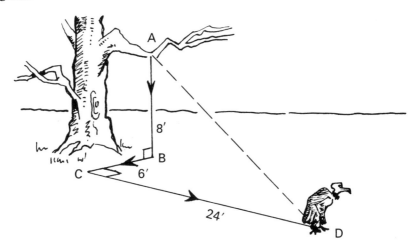

How far is the bird from its original position? (i.e., what is the length of the line AD?)

A. 26 ft.

B. 30 ft.

C. 34 ft.

D. 38 ft.

E. 42 ft.

44. What is the length of a chord which subtends an angle of 30° at the circumference of a circle of radius 3″?

A. 6″

B. $3\sqrt{3}''$

C. $3''$

D. $\sqrt{3}''$

E. $1\frac{1}{2}''$

45. Given, $\dfrac{2x-1}{y} = \dfrac{(2x-1)^2}{4x^2-1}$, what is y in terms of x?

   A. $y = 2x - 1$

   B. $y = 2x + 1$

   C. $y = (2x - 1)^2$

   D. $y = x$

   E. $y = \dfrac{1}{2x+1}$

46. Quadrilateral ABCD is such that $|AB| = |AD|$ and $D\hat{A}B = D\hat{B}C = 1$ right angle. If $|DC| = 9''$ and $|BC| = 3''$, what is $|AB|$?

   A. $3''$

   B. $4\cdot 24''$

   C. $6''$

   D. $8\cdot 48''$

   E. $9''$

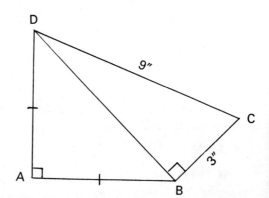

47. On a map of scale 1:2400 two points are $6''$ apart. If the points lie on the 1000' and 1500' contours respectively, what is their actual distance apart?

   A. $538\cdot 5'$

   B. $1,200'$

   C. $1,250'$

D. 1,300'

E. 1,700'

48. A solid brass cuboid is of length l, breadth b and height h. The brass is beaten into the shape of an open rectangular box of thickness t. If the internal dimensions of the box are length l, breadth b and depth h, which of the following is the correct equation in t?

   A. $(l + 2t)(b + 2t)(h + t) = 0$

   B. $(l + 2t)(b + 2t)(h + t) = lbh$

   C. $(l + 2t)(b + 2t)(h + t) = 2lbh$

   D. $lt + bt + ht = lbh$

   E. $t^3 = lbh$

49. If $\log X = 3(\log P + \log Q + \log R)$, which of the following is the correct relationship between X, P, Q and R?

   A. $X = 3 \times P \times Q \times R$

   B. $X = (P \times Q \times R)^3$

   C. $X = (P + Q + R)^3$

   D. $X = 3(P + Q + R)$

   E. $3X = P \times Q \times R$

50. A car is travelling at v m.p.h. in a town whose speed limit is 30 m.p.h. If the speed of the car is increased by 5 m.p.h. the speed limit will be exceeded. Which of the following statements represents this information?

   A. $v + 5 = 30$

   B. $v + 5 = 31$

   C. $v + 5 > 30$

   D. $v + 5 < 30$

   E. $v = 35$

Test 5
75 minutes 50 questions

1. In the diagram, , AB̂C = 41° and BD̂C = 63°. What is the size of CB̂D?

   A. 104°
   B. 90°
   C. 76°
   D. 63°
   E. 22°

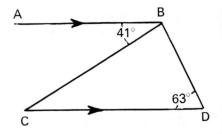

2. 7·988 x 10⁶ is equal to . . .

   A. 7,988,000,000
   B. 7,988,000
   C. 798,800
   D. 0·000007988
   E. 0·0000007988

3. 1621 was subtracted from 6244 and the result was 4323. In what base are all these numbers?

218

A. Five

B. Six

C. Seven

D. Ten

E. Twelve

4. Given, $p = \dfrac{ka^2}{b}$, what is k if $p = 4$ when $a = 100$ and $b = 12$?

   A.   0·0048

   B.   0·48

   C.   0·0003

   D.   0·03

   E.   4800

5. TAB and TC are tangents to the circle at A and C respectively. If $B\hat{A}C = 112°$, what is the size of $A\hat{T}C$?

   A. 68°

   B. 60°

   C. 58°

   D, 56°

   E. 44°

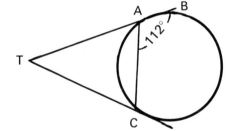

6. The line $4y + 2x = 3$ intersects the x-axis at A. What is the value of x at A?

   A. 0

   B. $\frac{1}{2}$

   C. $\frac{3}{4}$

   D. $1\frac{1}{2}$

   E. 2

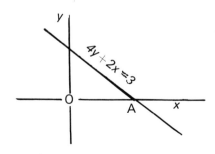

7. Simplify $(\sqrt{16})^2 + 27^{1/3} - 5^0$.

   A. 7

   B. 18

   C. 19

   D. 24

   E. 12

8. The angles of a triangle are in the ratio 3:4:2. What is the size of the smallest angle?

   A. 2°

   B. 20°

   C. 40°

   D. 60°

   E. 80°

9. Express $(x - 4)(5 - x) = 3$ in a form which can be more easily solved.

   A. $x^2 - 9x + 23 = 0$

   B. $x^2 + x + 23 = 0$

   C. $x^2 + 9x - 23 = 0$

   D. $x^2 + 9x + 17 = 0$

   E. $x^2 - 9x + 17 = 0$

10. ABCD is a parallelogram of area 72 sq. cm. E is a point on DC produced such that $\overline{AE}$ cuts $\overline{BC}$ at M where $|BM| : |MC| = 2 : 1$. What is the area of $\triangle CEM$?

    A. 24 sq. cm.

    B. 18 sq. cm.

    C. 12 sq. cm.

    D. 8 sq. cm.

    E. 6 sq. cm.

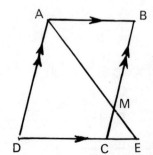

11. Use tables to find the square root of 0·00004567.

    A. 0·002137

    B. 0·006758

    C. 0·0002137

    D. 0·00002137

    E. 0·00006758

12. The following were given as factors of $50x^2 - 200y^2$:
    I. $(x - 2y)$ II. $(x + 2y)$ III. 50
    Which is (are) correct?

    A. I only

    B. II only

    C. III only

    D. I and III

    E. I, II and III

13. In $\triangle$ ABC, $\hat{B}$ is a right angle, |BC| = 2" and $A\hat{C}B$ = 60°. What is |AC|?

    A. $4\sqrt{3}"$

    B. $4"$

    C. $2\sqrt{3}"$

    D. $\frac{4}{\sqrt{3}}"$

    E. $1"$

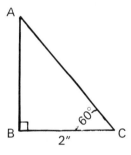

14. What is the sum of the roots of the equation, $x^2 - 6x = 0$?

    A. 0

    B. 6

    C. −6

    D. 1

    E −5

15. In the diagram, $\overline{TA}$ is a tangent to the circle ABC. If $T\hat{A}C = 66°$, what is the size of $A\hat{B}C$?

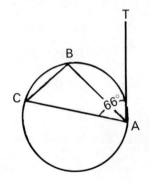

A. 114°

B. 104°

C. 66°

D. 90°

E. 132°

16. The lengths of the adjacent sides of a rectangle are P'' and Q''. If the length of each side is increased by 30%, by what percentage will the area increase?

A. 30%

B. 60%

C. 69%

D. 900%

E. Impossible unless we know the values of P and Q.

17. Which of the following can be factorized?

   I. $a^2 - b^2$ II. $c^2 - c - 36$ III. $d^2 - 18d + 81$

A. I only

B. II only

C. Both I and II

D. III only

E. Both I and III

18. How many sides has a regular polygon if each interior angle is 156°?

A. 12

B. 15

C. 24

D. 25

E. 30

19. The x and y axes on the given graph are drawn to the same scale. Which of the lines numbered *1, 2, 3, 4, 5* has the equation $y = x$?

    A. *1*
    B. *2*
    C. *3*
    D. *4*
    E. *5*

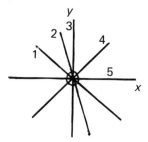

20. ABCD is a parallelogram such that $\overline{AD}$ is a tangent to the circle through B, C and D. If $D\hat{A}B = 73°$, what is the size of $B\hat{D}C$?

    A. 56°
    B. 44°
    C. 34°
    D. 17°
    E. More information is needed.

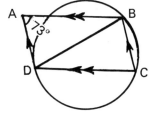

21. Take 1101_{two} from 110110_{two}.

    A. 101000_{two}
    B. 110111_{two}
    C. 1000011_{two}
    D. 101001_{two}
    E. 10_{two}

22. A car, whose wheels are of radius 14″, is travelling at 27 m.p.h. How many revolutions per minute is one of its wheels turning through? (Approximate π to $\frac{22}{7}$.)

A.   9 r.p.m.

B.   88 r.p.m.

C.   324 r.p.m.

D.   648 r.p.m.

E. 19,440 r.p.m.

23. If $a + b = 3a - 2b = 5$, what does a equal?

   A. $1\frac{1}{4}$

   B. 2

   C. $2\frac{1}{2}$

   D. 3

   E. 5

24. In the diagram, $P\hat{O}Q = 150°$ and the radius of the circle, $\overline{OP}$, is 4·2" in length. What is the length of arc PRQ? (Take π to be $\frac{22}{7}$.)

   A. 11"

   B. 15·4"

   C. 17·64"

   D. 23·1"

   E. 32·34"

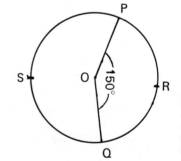

25. . . . and what is the area of sector OPSQ?

   A. 11 sq. in.

   B. 15·4 sq. in.

   C. 17·64 sq. in.

   D. 23·1 sq. in.

   E. 32·34 sq. in.

26. What is the range of values of x for all points which lie within the unshaded areas of the graph?

   A. $-5 > x > 2$
   B. $-5 < x < 2$
   C. $-5 < x > 2$
   D. $-5 > x < 2$
   E. $-5 = x = 2$

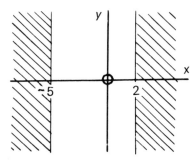

27. In the diagram, O is the centre of the circle. Chord AB ∥ tangent XY and OAX and OBY are straight lines. If $|OB| = 5''$ and $|AB| = 6''$, what is $|AX|$?

   A. $1''$
   B. $1\frac{1}{4}''$
   C. $\sqrt{2}''$
   D. $1\frac{9}{16}''$
   E. $2''$

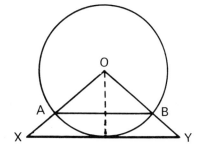

28. Two planes fly a distance of 500 miles. The average speed of the faster plane is 40 m.p.h. more than the average speed of the slower plane and it takes 40 minutes less for the journey. If x m.p.h. is the average speed of the *slower* plane, which of the following equations is correct?

   A. $\dfrac{500}{x+40} - \dfrac{500}{x} = 40$

   B. $\dfrac{500}{x+40} - \dfrac{500}{x} = \dfrac{2}{3}$

   C. $\dfrac{500}{x} - \dfrac{500}{x+40} = 40$

   D. $\dfrac{500}{x} - \dfrac{500}{x+40} = \dfrac{2}{3}$

   E. $500(2x + 40) = \dfrac{2}{3}$

225

29. Simplify the ratio, $\dfrac{1}{\sqrt{2}} : \sqrt{50}$.

   A. 1:100

   B. 1:20

   C. 1:10

   D. 1:5

   E. 2:5

30. Which of the following equations has $-\frac{1}{2}$ and $\frac{3}{4}$ as its roots?

   A. $8x^2 - 2x - 3 = 0$

   B. $8x^2 + 2x - 3 = 0$

   C. $8x^2 - 10x + 3 = 0$

   D. $8x^2 + 10x - 3 = 0$

   E. $4x^2 + 2x - 3 = 0$

31. 'The sum of the angles . . . is two right angles.' Which of the following does *not* complete the statement?

   A. . . . in opposite segments of the same circle . . .

   B. . . . in a triangle . . .

   C. . . . on a straight line . . .

   D. . . . at a point . . .

   E. . . . in half a revolution . . .

Use the following diagram and information to answer questions 32 and 33.

The diagram is a cross-sectional view of a swimming pool and diving board. The width of the pool is 30 feet; all the other dimensions are on the diagram.

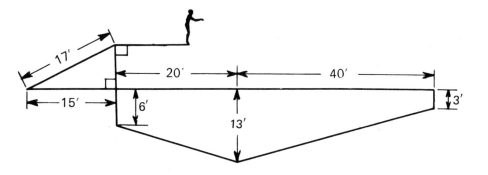

32. What is the volume of water in the pool?

    A. 2,700 cub. ft.

    B. 5,100 cub. ft.

    C. 8,100 cub. ft.

    D. 15,300 cub. ft.

    E. 30,000 cub. ft.

33. If the man standing on the diving board is 6 feet tall, what is the height of the top of his head above the lowest point of the pool?

    A. 19 ft.

    B. 27 ft.

    C. 36 ft.

    D. $41\frac{2}{3}$ ft.

    E. Impossible, unless we are told the length of the diving board.

34. Two towns on level ground are 20 miles apart. On a map of the locality the towns are $26\frac{2}{3}$ inches apart. What is the scale of the map?

    A. 1:20

    B. $1:1,333\frac{1}{3}$

    C. 1:9,600

    D. 1:16,000

    E. 1:48,000

35. $x = y \times z$. If x, y and z are whole numbers, which of the following statements is (are) true?

   I. If y and z are even, x must be even.
   II. If x and y are even, z must be even.
   III. If x is even and y is odd, z must be even.

   A. I only

   B. II only

   C. III only

   D. Both I and II

   E. Both I and III

36. ABCD is a kite in which $|AB| = 3"$, $|BC| = 4"$ and $|AC| = 5"$. What is $|BD|$?

   A. 5"

   B. 4·8"

   C. 4"

   D. 3·2"

   E. 2·4"

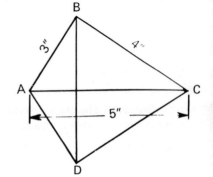

37. Factorize $2x^2 - x - 2ax + a$.

   A. $(x - a)(2x - 1)$

   B. $(x + a)(2x - 1)$

   C. $2x(x - a)$

   D. $(x + 1)(2x - a)$

   E. $2x(x - a + 1)$

38. A car covers the first 80 miles of a journey in 2 hours and then completes the journey by travelling for a further $2\frac{1}{2}$ hours at 50 m.p.h. What is its average speed for the whole journey?

   A. 45 m.p.h.

B. $47\frac{1}{4}$ m.p.h.

C. $63\frac{1}{3}$ m.p.h.

D. $44\frac{4}{9}$ m.p.h.

E. $45\frac{5}{9}$ m.p.h.

39. $\overline{PM}$ and $\overline{PN}$ are tangents to the circle, centre O. If $M\hat{N}O = x°$, express $M\hat{P}N$ in terms of x.

    A. $2x°$
    B. $x°$
    C. $(90 - x)°$
    D. $(180 - 2x)°$
    E. $(90 - 2x)°$

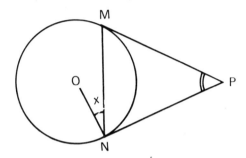

40. Use tables to find the reciprocal of $1·168 \times 10^7$.

    A. $8·682 \times 10^{-7}$
    B. $8·682 \times 10^{-8}$
    C. $8·560 \times 10^7$
    D. $8·560 \times 10^{-7}$
    E. $8·560 \times 10^{-8}$

41. For what values of x is the fraction $\dfrac{x+1}{x^2 + x - 12}$ undefined?

    A. $-1, +4$ and -3
    B. $-1, -4$ and $+3$
    C. $+4$ and -3 only
    D. -4 and $+3$ only
    E. $+12$ only

42. Accompanying each of the given sketch graphs is a statement about the graph. Which graph is coupled with an incorrect statement?

A.
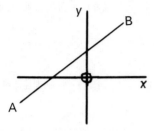
AB has a positive gradient.

B.
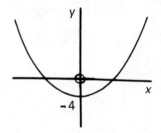
The equation of the curve is $y = x^2 - 4$.

C.
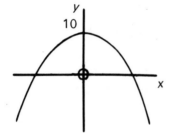
The equation of the curve is $y = x^2 + 10$.

D.
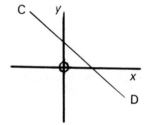
CD has a negative gradient.

E.
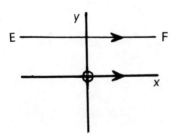
The gradient of EF is zero.

43. A rectangular window contains n panes of glass. How many identical panes would be contained by a window which is twice as long and twice as wide?

   A. $2n$

   B. $4n$

   C. n^2

   D. $2n^2$

   E. $4n^2$

44. A solid is made, as in the diagram, by combining a hemisphere and a cylinder of equal radii. If this radius is 3 in. and the height of the cylinder is 5 in., what, in terms of π, is the volume of the solid? (The volume of a sphere, radius r, is $\frac{4}{3}\pi r^3$.)

   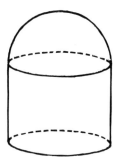

   A. 63π cub. in.

   B. 81π cub. in.

   C. 108π cub. in.

   D. 54π cub. in.

   E. 48π cub. in.

45. What is the product of 24_{eight} and 5_{eight}?

   A. 100_{eight}

   B. 120_{eight}

   C. 124_{eight}

   D. 144_{eight}

   E. 150_{eight}

46. The graphs of $y = x^2 + 3x + 1$ and $y = 2x^2 - 4x + 2$ intersect at two points. Which of the following equations will have as its solutions the values of x at these points?

   A. $x^2 + 3x - 1 = 0$

   B. $3x^2 - x + 3 = 0$

C. $x^2 - 7x + 1 = 0$

D. $x^2 + 7x - 1 = 0$

E. $2x^2 - 4x + 1 = 0$

47. A crescent is formed, as in the diagram, by drawing a circle within a larger circle. The common chord, XY, subtends an angle of 60° at A, the centre of the large circle, and an angle of 90° at B, the centre of the small circle. If $|XY| = 2''$, what is $|AB|$?

A. $1''$

B. $(2 - \sqrt{3})''$

C. $(\sqrt{3} - \sqrt{2})''$

D. $(\sqrt{3} - 1)''$

E. $(2 - \sqrt{2})''$

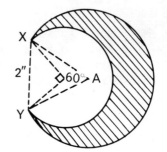

48. The lengths of the perimeters of a square of side Y'' and a circle of radius X'' are equal. What is X in terms of Y?

A. $X = \dfrac{4Y}{\pi}$

B. $X = \dfrac{2Y}{\pi}$

C. $X = \dfrac{Y}{\sqrt{\pi}}$

D. $X = \dfrac{Y}{\pi}$

E. $X = \dfrac{\pi Y}{2}$

49. M is the point of intersection of the diagonals of square ABCD. Points P and Q are taken on $\overline{AB}$ and $\overline{AD}$ so that $|AP| = 3''$ and $|AQ| = 5''$. If $|BC| = 8''$, what is the area of quadrilateral APMQ?

A. 15 sq. in.

B. 16 sq. in.

C. 18 sq. in.

D. 24 sq. in.

E. 30 sq. in.

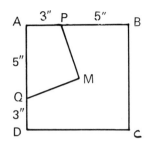

50. In the diagram, |DA| = |DB| = |DC|, BÂC = 60° and AD̂C = 140°. What is the size of AĈB?

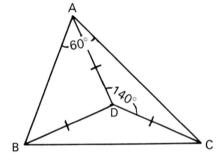

A. 50°

B. 60°

C. 70°

D. 75°

E. 80°

Test 6
75 minutes 50 questions

1. A cistern has a capacity of 40 pints and is filled with water which weighs 1·25 lb. per pint. What is the weight of water in the cistern?

   A. 0·03125 lb.

   B. 0·5 lb.

   C. 31·25 lb.

   D. 32 lb.

   E. 50 lb.

2. A weaver bought a bundle of grass for 5/-, from which he made 8 mats. If he sold the mats for 1/6d. each, what was his percentage profit?

   A. 240%

   B. 140%

   C. 120%

   D. 40%

   E. 20%

3. In the diagram, $\overline{AB} \parallel \overline{CD}$ and $\overline{AD}$ and $\overline{BC}$ intersect at E. If $A\hat{E}C = 3x°$ and $E\hat{C}D = x°$, what is the size of $E\hat{A}B$?

234

A. $x°$

B. $2x°$

C. $3x°$

D. $180° - 3x°$

E. $180° - 4x°$

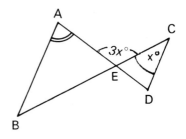

4. When revolving at its minimum, a fan does 80 r.p.m. At its maximum, its speed is increased in the ratio 9:2. What is its maximum speed?

   A. 320 r.p.m.

   B. 350 r.p.m.

   C. 360 r.p.m.

   D. 450 r.p.m.

   E. 720 r.p.m.

5. In the diagram, O is the centre of the circle, $|BD| = |BC|$ and $A\hat{B}C = 68°$. What is the size of $A\hat{C}D$?

   A. 34°

   B. 45°

   C. 56°

   D. 62°

   E. 68°

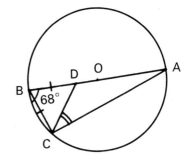

6. A woman buys 3 large eggs at 4d. each and 9 small eggs at 2d. each. What is the average cost of an egg?

   A. $3\frac{1}{2}$d.

   B. 3d.

   C. $2\frac{2}{3}$d.

   D. $2\frac{1}{2}$d.

   E. $2\frac{1}{3}$d.

235

7. If 1 cub. cm. is approximately equal to 0·061024 cub. in., what is 1 litre in cub. in.?

   A.   0·000061024 cub. in.

   B.   0·00061024 cub. in.

   C.   0·61024 cub. in.

   D.   6·1024 cub. in.

   E.   61·024 cub. in.

8. If $5x + 3y = 4$ and $5x - 3y = 2$, what is the value of $25x^2 - 9y^2$?

   A.   8

   B.   12

   C.   16

   D.   20

   E.   64

9. A man deposited £160 in a savings account which paid simple interest at the rate of 5% per annum. After a length of time, he withdrew all his money at once; this amounted to £178. How long had he left his money in the account?

   A. $5\frac{1}{3}$ mon.

   B.   2 yr. 3 mon.

   C.   2 yr. 6 mon.

   D.   4 yr. 6 mon.

   E.   22 yr. 6 mon.

10. What is the value of $5b + (a + b)^2$ if $a = 3$ and $b = -7$?

    A.   65

    B.   51

    C.   19

    D.   −19

    E.   −51

11. Simplify $\dfrac{3}{\sqrt{2}} - \dfrac{1}{2\sqrt{2}}$.

   A. $2\tfrac{1}{2}$

   B. $1\tfrac{1}{4}$

   C. $\tfrac{3}{4}$

   D. $\dfrac{5\sqrt{2}}{4}$

   E. $\dfrac{\sqrt{2}}{2}$

12. Two men are standing a distance d feet apart on the same path. At the same instant they start running along the path in the same direction. To begin with, the faster runner, whose speed is P ft./sec. is behind the slower runner, who does Q ft./sec. How many seconds is it until the faster man catches up the slower man?

   A. $\dfrac{d}{P+Q}$

   B. $d(P+Q)$

   C. $\dfrac{d}{P-Q}$

   D. $d(P-Q)$

   E. $\dfrac{P-Q}{d}$

13. Express $\dfrac{1}{a-b} - \dfrac{1}{a+b}$ as a single fraction.

   A. $\dfrac{1}{2a}$

   B. $\dfrac{-1}{2b}$

   C. $\dfrac{2a}{a^2 - b^2}$

237

D. $\dfrac{2b}{a^2 - b^2}$

E. $\dfrac{2}{a - b}$

14. If $(4x + 1)^2 = 16x^2 + 25$, what is the value of x?

    A. −3

    B. 1

    C. 3

    D. 4

    E. x cannot be determined from an equation of this type.

15. ABCDE is a pentagon such the |AE| = |AD| = |ED| = |BD| = |BC| = |CD|. If $A\hat{B}D = 70°$, what is the size of the largest interior angle of the pentagon?

    A. 160°

    B. 140°

    C. 130°

    D. 120°

    E. 100°

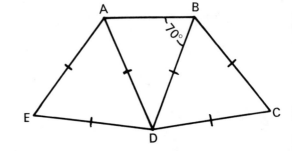

16. Quadrilateral ABCD is drawn within a rectangle PQRS so that $\overline{AB}$, $\overline{BC}$, $\overline{CD}$ and $\overline{DA}$ are each 1″ from $\overline{PQ}$, $\overline{QR}$, $\overline{RS}$ and $\overline{SP}$ respectively.
    Three boys are talking about the diagram:
    Albert: 'ABCD must be a rectangle.'
    Kwesi: 'Yes, and it is similar to rectangle PQRS.'
    Mensah: 'Points A, B, C, D are exactly $\sqrt{2}$″ from points P, Q, R, S respectively.'
    Which boy(s) is (are) correct?

    A. Albert only

    B. Albert and Kwesi only

C. Kwesi only

D. Albert and Mensah only

E. All the boys are correct.

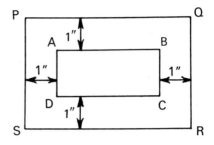

17. If $81x^2 + ax - 4 = (9x + 2)(9x - 2)$, what is the value of a?

    A. -36

    B. -18

    C. 0

    D. 18

    E. 36

18. If a varies directly as b^2 and inversely as c, then b varies as . . .

    A. $\sqrt{\dfrac{a}{c}}$

    B. ac

    C. $\left(\dfrac{a}{c}\right)^2$

    D. $(ac)^2$

    E. $\sqrt{ac}$

19. In △ ABC, |AB| = 5", |AC| = 3" and |BC| = 4". $\overline{CD}$ is the perpendicular from C to $\overline{AB}$. What is |CD|?

    A. 2·5"

    B. 2·4"

    C. 2·2"

    D. 2·0"

    E. 1·8"

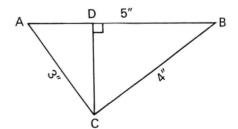

20. Evaluate $(-32)^{2/5}$.

    A. $-12 \cdot 8$
    B. $-\frac{1}{4}$
    C. $\frac{1}{4}$
    D. -4
    E. 4

21. In the diagram, O is the centre of the circle and $\overline{TA}$ is a tangent to the circle at A. If $B\hat{A}T = x$, what is $B\hat{O}A$ in terms of x?

    A. $2x$
    B. $180° - 2x$
    C. x
    D. $180° - x$
    E. $90° - x$

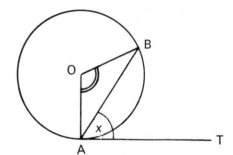

22. $\overline{DX}$ and $\overline{EX}$ are the perpendicular bisectors of the sides AB and AC respectively of $\triangle ABC$. If $B\hat{X}C = 126°$ and $A\hat{B}C = 48°$, what is the size of $A\hat{C}B$?

    A. $69°$
    B. $66°$
    C. $64°$
    D. $63°$
    E. $54°$

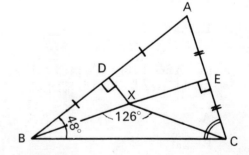

23. An office has fifteen '40 watt' light bulbs. How many '25 watt' light bulbs could it have, it the total wattage was the same in each case?

    A. 9

B. 15

C. 24

D. 25

E. 40

24. △ABC is equilateral and D is the mid-point of $\overline{AC}$. What is tan $A\hat{B}D$?

   A. $\frac{1}{2}$

   B. $\sqrt{3}$

   C. $\frac{\sqrt{3}}{2}$

   D. $\frac{1}{\sqrt{3}}$

   E. $\frac{2}{\sqrt{3}}$

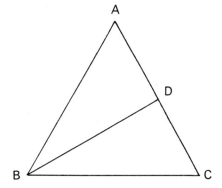

25. △ABC is right angled at C, |AB| = 10″ and $A\hat{B}C = 41°$. Use this information to find an expression for the area of the triangle.

   A. 25 sq. in.

   B. 50 sq. in.

   C. 30 sin 41° sq. in.

   D. 50 sin 41° sq. in.

   E. 50 sin 41° cos 41° sq. in.

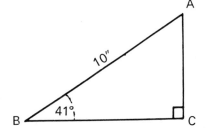

26. The exterior angles of a hexagon are in the ratio 3:3:3:4:5:6. What is the size of the largest interior angle of the hexagon?

   A. 145°

   B. 135°

   C. 105°

   D. 90°

   E. 75°

27. $\overline{TA}$ and $\overline{TB}$ are tangents to the circle ABC at A and B. $\overline{TA} \parallel \overline{BC}$ and $A\hat{T}B = 100°$. What is the size of $C\hat{A}B$?

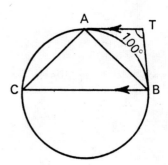

A. 130°
B. 100°
C. 90°
D. 85°
E. 80°

28. If $T = \dfrac{2\pi}{\omega}$ and $\omega^2 = \dfrac{g}{l}$, express T in terms of π, g and l.

A. $T = \dfrac{2\pi g}{l}$

B. $T = \dfrac{2\pi l}{g}$

C. $T = 2\pi\sqrt{\dfrac{g}{l}}$

D. $T = 2\pi\sqrt{\dfrac{l}{g}}$

E. $T = \dfrac{2\pi l^2}{g^2}$

29. Express 0·00529 in the form $P \times 10^a$, where P lies between 1 and 10 and a is a positive or negative integer.

A. $5·29 \times 10^3$
B. $5·29 \times 10^2$
C. 529×10^5
D. $5·29 \times 10^{-2}$
E. $5·29 \times 10^{-3}$

30. ABCD is a parallelogram. Points K, L, M, N are taken on $\overline{AB}$, $\overline{BC}$, $\overline{CD}$, $\overline{DA}$ respectively so that $\frac{|BK|}{|KA|} = \frac{|BL|}{|LC|} = \frac{|CM|}{|MD|} = \frac{|AN|}{|ND|} = \frac{1}{2}$. Which of the following describes quadrilateral KLMN?

A. Parallelogram

B. Rhombus

C. Kite

D. Trapezium

E. Cyclic Quadrilateral

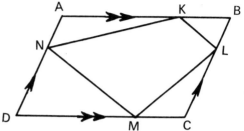

31. A circle and rectangle PQRS have the same area A. If the length of a diameter of the circle is equal to $|PQ|$, what, in terms of A and π, is $|QR|$?

A. $\dfrac{\pi}{2A}$

B. $\tfrac{1}{2}\sqrt{\pi A}$

C. $\dfrac{A(\pi - 1)}{2\pi}$

D. $\sqrt{\pi A}$

E. $\tfrac{1}{2}\sqrt{\dfrac{A}{\pi}}$

32. The locus of points in space equidistant from a given straight line is

A. a pair of parallel straight lines.

B. a plane, perpendicularly bisecting the given line.

C. a cylinderical surface.

D. a circle.

E. a spherical surface.

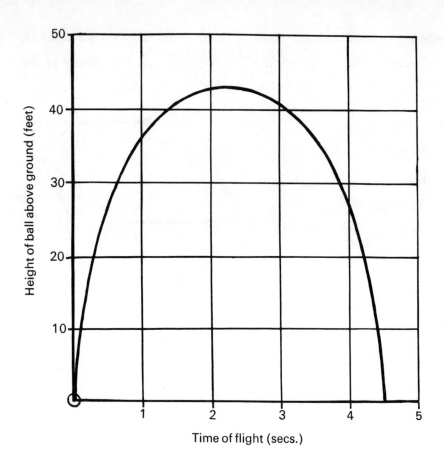

Use this graph to answer questions 33, 34, 35, 36.

The graph shows how the height of a ball, which has been thrown upwards, varies with time during its flight.

33. After how many seconds is the ball at its highest point?

   A. 2 secs.

   B. 2·25 secs.

   C. 2·5 secs.

   D. 4·5 secs.

   E. 43 secs.

34. What is the greatest height reached by the ball?

    A. 4·5 ft.

    B. 40 ft.

    C. 43 ft.

    D. 46 ft.

    E. 50 ft.

35. At what height is the ball after 3·7 seconds?

    A. 334 ft.

    B. 35 ft.

    C. 37 ft.

    D. 39 ft.

    E. 42 ft.

36. During its flight, the ball is at a height of 23 feet on two occasions. What is the length of time between those two occasions?

    A. 4·1 secs.

    B. 0·4 secs.

    C. 3·9 secs.

    D. 3·7 secs.

    E. 4·2 secs.

37. What is the range of values of x for which $2x + 3 > 3x - 12$?

    A. $x < 3$

    B. $x > 3$

    C. $x > 15$

    D. $x > -15$

    E. $x < 15$

38. In the diagram, $\overline{DE} \parallel \overline{BC}$, $|DE| = 2''$ and $|BC| = 6''$. If the area of △ADE is 2 sq. in., what is the area of trapezium DECB?

A. 16 sq. in.
B. 12 sq. in.
C. 8 sq. in.
D. 7 sq. in.
E. 4 sq. in.

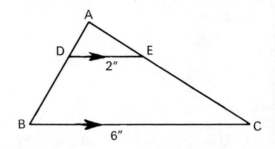

39. In the diagram, $\overline{AD}$ is the perpendicular from A to $\overline{BC}$, $|ED| = 1''$ and $|AD| = |CD| = \sqrt{3}''$. What is the size of angle BAC?

A. 135°
B. 105°
C. 90°
D. 75°
E. 60°

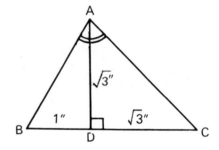

40. The diagram shows a metal plate, right-angled at each corner. If the metal is melted down and recast as a square plate of the same thickness, what would be the perimeter of the square?

A. 52"
B. 48"
C. 46"
D. 44"
E. 40"

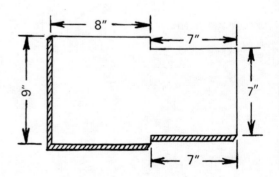

41. A cylinder, 12″ high, contains water to a depth of 8″. A cone, 6″ high, whose base diameter is equal to that of the cylinder is immersed in the water. What is the new depth of water in the cylinder?

   A. 12″

   B. 11″

   C. 10″

   D. 9½″

   E. 9″

42. The inscribed circle of right-angled triangle PQR touches the hypotenuse PR at T. If |PT| = 2 cm. and |TR| = 3 cm., what is the length of the radius of the in-circle?

   A. 1 cm.

   B. 1¼ cm.

   C. 1½ cm.

   D. 2 cm.

   E. 2½ cm.

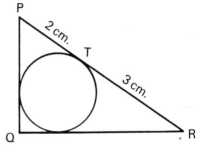

43. If 1 inch is approximately equivalent to 2·5 cm., how many yards is 1 metre approximately equivalent to?

   A. 0·9 yd.

   B. 1·0 yd.

   C. 2·5 yd.

   D. 1·1 yd.

   E. 1·25 yd.

44. A radio mast, 150 feet high, is supported by a wire stretched between the top of the mast and a point on the horizontal ground. The mast is vertical and it makes an angle of 38° with the wire. What is the length of the wire?

   A. 150 sec 38° ft.

B. 150 cosec 38° ft.

C. 150 cot 38° ft.

D. 150 cos 38° ft.

E. 150 tan 38° ft.

45. Which of the following is a factor of $(a + b)^2 + a^2 - b^2$?

   A. $2a$

   B. $(a - b)$

   C. $(a + 2b)$

   D. $2b$

   E. None of the above.

46. D, E, F are the mid-points of the sides AB, BC, CA of △ABC. What is the ratio of the area of △DEF to the area of △ABC?

   A. Impossible, unless we know the lengths of the sides of △ABC.

   B. 2:3

   C. 1:2

   D. 1:3

   E. 1:4

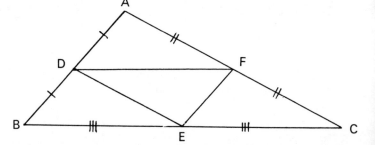

47. In the diagram, O is the centre of circle ADC, and AOB and DCB are straight lines. $A\hat{B}D = x$ and $|CO| = |CB|$. What is the size of $A\hat{O}D$ in terms of x?

   A. $5x$

   B. $4x$

   C. $3x$

   D. $2x$

   E. $180° - 4x$

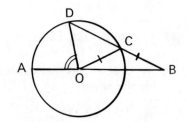

48. Use the measurements given in the diagram to find an expression for |PQ|.

A. $\dfrac{5 \sin 73°}{\sin 46°}$

B. $\dfrac{5 \sin 73°}{\sin 27°}$

C. $\dfrac{5 \sin 27°}{\sin 73°}$

D. $\dfrac{5 \sin 46°}{\sin 27°}$

E. $\dfrac{5 \sin 27°}{\sin 46°}$

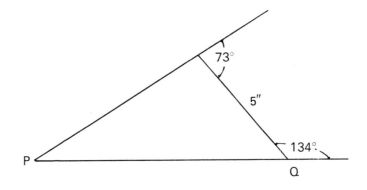

49. If the equations, $x^2 - 5x + 6 = 0$ and $x^2 + ax + 2 = 0$, have a common root, what is the value of a?

   A. +3
   B. +2
   C. −1
   D. −2
   E. −3

50. The circumference of a circle is 14π units in length. What would be the increase in length of the circumference if the radius of the circle was increased by 1 unit?

   A. 1 unit
   B. 2 units
   C. π units
   D. 2π units
   E. 7π units